SPACE, TIME AND CAUSALITY

SPACE, TIME AND CAUSALITY

An Essay in Natural Philosophy

J. R. LUCAS

Fellow of Merton College, Oxford

AT THE CLARENDON PRESS
1984

Oxford University Press, Walton Street, Oxford OX2 6DP
London New York Toronto
Delhi Bombay Calcutta Madras Karachi
Kuala Lumpur Singapore Hong Kong Tokyo
Nairobi Dar es Salaam Cape Town
Melbourne Auckland
and associated companies in
Beirut Berlin Ibadan Mexico City Nicosia

Oxford is a trade mark of Oxford University Press

Published in the United States
by Oxford University Press, New York

British Library Cataloguing in Publication Data
Lucas, J. R.
Space, time and causality.
1. Physics—Philosophy
I. Title
530′.01   QC6
ISBN 0-19-875057-9
ISBN 0-19-875058-7 Pbk

Library of Congress Cataloging in Publication Data
Lucas, J. R. (John Randolph), 1929–
Space, time, and causality.
Bibliography: p.
Includes index.
1. Philosophy—Addresses, essays, lectures.   2. Physics—
Philosophy—Addresses, essays, lectures.   I. Title.
B29.L72   1984   113   84-9681
ISBN 0-19-875057-9
ISBN 0-19-875058-7 (pbk.)

Printed in Great Britain
at the University Press, Oxford
by David Stanford
Printer to the University

*To Eric James
and all my teachers*

Preface

THIS work is for the most part a write-up of lectures given in the 1970s to first-year men reading Physics and Philosophy or Mathematics and Philosophy. Their set texts are: David Hume, *An Enquiry Concerning Human Understanding;* H. G. Alexander, *The Leibniz–Clarke Correspondence*, Manchester, 1956; and Bertrand Russell, *An Introduction to Mathematical Philosophy*, London, 1919. I have paid particular attention to themes arising from the first two of these. I have also shown, sometimes with wearisome explicitness, the applications of formal logic to the formulation of issues and the elucidation of problems. No special knowledge of Relativity Theory or Quantum Mechanics is called for, although occasional indications are given of problems which they generate but which lie beyond the horizon of this work.

Although the text is self-contained and can be read as a book, I have retained occasional interjections as exercises or as giving food for further thought. I have also retained, lists of preliminary and further reading. Philosophy is an activity rather than a set of truths, and the reader may benefit more by working his way through the preliminary reading before attempting each chapter, than by reading straight through the book itself. I have put letters A, B and C against the preliminary reading and sometimes the further reading. The items marked A should not take more than one hour to read, and offer a lead into the main ideas being discussed. The items marked B are well worth reading, but are partly covered by the text. The items marked C open up new issues, and help the reader to develop his own ideas, but are not integral to the theme of this book. There are, of course, many other important works not recommended here—they can be found in *A Study Guide to the Philosophy of Physics*, by W. H. Newton-Smith: but time is short, and the bibliographical help offered in this book seeks to be helpful by being very selective.

The book owes a lot to W. H. Newton-Smith, Fellow of Balliol, H. R. Harré, Fellow of Linacre, my colleagues C. J. H. Watson, D. Bostock, Fellows of Merton, P. E. Hodgson, Fellow of Corpus Christi, and I. J. R. Aitchison, Fellow of Worcester.

In the fullness of time I envisage there being further volumes, perhaps by other hands, dealing with the philosophy of the Special and the General Theory of Relativity, and with the philosophy of Quantum Mechanics.

Merton College, Oxford   J. R. L.
October, 1983

References

PARAGRAPH references to David Hume's *An Enquiry Concerning Human Understanding* and page references to both that and his *Treatise on Human Nature* are to the editions edited by L. A. Selby-Bigge, and published by the Oxford University Press. Page references to the Leibniz–Clarke Correspondence, together with some selections from Newton's *Principia*, are to the edition edited by H. G. Alexander, and published by Manchester University Press. The *Feynman Lectures* are by Richard P. Feynman, Robert B. Leighton and Matthew Sands, published by Addison-Wesley, 1963.

Contents

I	The Principles of Natural Philosophy	1
II	Induction	14
III	Causality	27
IV	The Logic of Cause and Effect	44
V	"Contiuity" and the *A Priori*	69
VI	Measurement	83
VII	Constancy, Invariance, and Symmetry	105
VIII	Homogeneity and Isotropy	119
IX	Reflections, Relationism, and Parity	143
X	Causal Cords	156
XI	Fields	176
XII	Continuity and Beyond	184
	Appendix on Relationism	191
	Index	197

The Principles of Natural Philosophy

OUR understanding of natural phenomena has been much mis-understood. Plato, and with him most of the Greek philosophers, set out to achieve an entirely deductive science of Nature. They sought some first principle from which all others could be deduced. Recognising that there was some pattern of events, and believing the universe to be fundamentally orderly and rational, they sought to discover by ratiocination alone the nature of things, and by cogitating on the concept of harmony they hoped to determine what Nature must, and therefore would, be like. This was wrong. It was wrong because it failed to account for the element of sheer brute fact which constitutes part of our experience of Nature. Things are what they are, and no amount of arguing will make them otherwise. To put it another way, we can see that we cannot give a completely deductive account of Nature, because if the account is to be of any use it must enable us to derive certain conclusions about particular matters of fact, and these particular matters of fact could always be otherwise. We could make a film of our life, including all the experiments we did and all the observations that we ever made. We could also fake a second film in which the same experiments appear to be undertaken, but with results clean contrary to what science would lead us to expect. Our experience could thus be different from what in fact it is. It could correspond to the second film rather than the first. The value of science and of our common-sense beliefs about the world is that they lead us to believe that our experience always has corresponded, and always will, to the first film rather than the second. But in so far as it is valuable it exposes itself to the possibility of being wrong. By usefully telling us that it is like the first film, it is denying that it is like the second, and if our experience should follow the second film rather than the first, then our science and common sense would have been proved to be wrong. Thus there cannot be a completely deductive science of the natural world. It must be empirical at least in part, because if it is to be about the natural world, it must be about our experience of the natural world, and the only absolute proof that experience is as it was said to have been going to be is that it has actually occurred, and was as it was said to have been going to be. We can see this more clearly by following out the consequences of Plato's own approach. In the seventh book of the *Republic* (VII, 529) he dismisses altogether the relevance of empirical evidence, and laughs at those would-be scientists who waste their time lying flat on their backs observing celestial phenomena instead of getting down to the serious business of calculation. There is a similar story of the lady who boasted to Mrs Einstein of the extremely expensive telescope her husband used to find out the nature of the universe. "My husband does it," was the reply, "on the back of an old envelope." Nevertheless, we need to press the question what view we should take if the experimentalist, money-spending and banausic though he be, disagrees with the theoretician. What happens if the celestial bodies are not where, according to our celestial mechanics, they ought to be? Plato's answer is clear: "So much the worse for the celestial bodies. It is just what one might have expected with mere appearances, and just goes to show how unreliable they are, and how little use they are as guides to the true reality". It is a logically tenable position. It was taken up by the Rabbinical authorities of Judaism, who adopted certain calculations to determine when the Paschal full moon would occur, and found that the actual satellite of the earth did not always obey their ruling. No matter. They kept their ruling inviolate, and to this day the Jewish Passover, and therefore also the Christian Easter, is calculated according to the Golden Numbers which boys read at the beginning of their Prayer Books during dull sermons, and it sometimes is and sometimes is not the case that the calculated date coincides with the actual moon's being full for the first time after the spring equinox. But although it is a tenable position, it is an empty one. It divorces the Paschal full moon from the moon we can actually observe, and leaves it an entirely vacuous concept. If I cannot rely on the Paschal full moon being in any way apparent—if I cannot rely on it to light me home on my journey after dark—what use is it, and why should it be called a moon at all? Although, as we shall see, we are not absolutely obliged to take all appearances at face value, we cannot be so cavalier as to dismiss them altogether. Any scientific theory must at least try "to save the appearances" σώζειν τὰ φαινόμενα or it ceases to be science at all, and becomes just an uninterpreted calculus of pure mathematics. We reject Plato's extreme rationalist approach as being too *a priori*, and note as a further characteristic feature of physics that it is, in some sense, contingent.

Things do not have to be the way they are. It is always possible—logically possible, that is, not physically possible—that our theory may not be borne out by empirical evidence, and if so, it is the theory, not the empirical evidence, that has to go.

This fundamental insight of the contingency of physics has given rise to the empiricist philosophy of science, set up in self-conscious opposition to the extreme *a priorism* of Plato and the Rationalists. It takes many forms. At its mildest it simply insists upon the relevance of empirical evidence to our theories about the natural world; often, however, a much more radical empiricism is maintained, which, starting from a critique of the *a priori* theorizing of armchair philosophers, reaches the conclusion that science is simply an economical description of actual or possible empirical observations. The argument can be made most appealing by being put in almost tautological form. Some statements are "analytically" true, true in virtue of the meaning of their terms. For example, the statement that either it is raining or it is not raining must be true, whether it is raining or not, solely because of the meaning of the words 'either', 'or' and 'not'. There is no way in which the statement that either it is raining or it is not raining could be false; to contradict it would be to make a self-contradictory statement. But for the very reason that there is no way in which that statement could be false, there is no way in which it says anything about the world. Whatever the state of the weather, it is true; and therefore having asserted it, one has asserted nothing. An analytic statement commits one to nothing, and therefore one achieves nothing by committing oneself to it. Nothing has been said, because nothing has been ruled out.

Synthetic statements are statements which are not analytic, that is, statements whose contradictories are not self-contradictory. They do say something; and by the same token, it is logically possible for them to be wrong. We are interested in synthetic statements, because we do want to say something about the natural world, and want to be able to obtain and to give guidance about the course of natural events. But if we assert synthetic statements, we cannot justify them by showing them to be analytic, nor can we justify them by deducing them from other statements themselves analytic. For in either case we should have "justified" our synthetic statements only by showing them not to be synthetic after all, not capable therefore of doing the work for which we wanted them. And if to justify is to make useless, then justification itself is a useless activity. These and similar considerations have led philosophers to conclude that in some important sense there cannot be "synthetic propositions true *a priori*". If a proposition is to be genuinely synthetic, there must be a logically possible combination of sense experiences which would show that proposition to be false; that is, we cannot rule out *a priori* the possibility of its being false, and so we cannot absolutely prove *a priori* that it must be true.

This argument is not just an argument, but represents the tenor of a whole philosophical movement. The radical empiricists have been so much impressed with the insight that our theories about the world must be based on, and can be refuted by, experience, that they have taken experience to be the sole substantial ingredient of knowledge. Apart from the purely tautologous and vacuous reformulations of deductive logic, all knowledge and understanding is, they have claimed, experience and nothing but experience. In particular, recent exponents of radical empiricist views have reduced the theories of modern science to being simply useful and compendious modes of making predictions about future sense experience. Just that and nothing more.

Such a claim is intuitively implausible. We do not think of scientific theories as simply useful devices for predicting events. However great store we set by predictive power, we think prediction important in scientific theories as being a test of truth, rather than regarding truth as useful and as meaning nothing more than predictive power. As an account of what scientists take their theories to be, the empiricist account is simply false. Nor does it accord better with what scientists do than with what they think they do. The best example is the Special Theory of Relativity. This was originally based on the negative result of the Michelson–Morley experiment in 1887.[1] Einstein's theory could accommodate this result, which he cites in its support; after much controversy Einstein's theory was generally accepted by physicists, and many of them urged the negative result of the Michelson–Morley experiment as one of the crucial pieces of evidence in its favour. In the years from 1902 to 1926 the Michelson–Morley experiment was repeated many times and yielded a positive result.[2] Nobody, however, was prepared to reject the Theory of Special Relativity on this account. *Ad hoc* hypotheses of experimental error

[1] A. A. Michelson and F. W. Morley, *American Journal of Science*, 34, 1887, p. 333; *Phil. Mag.*; 24, 1887, p. 449.

[2] For full account see Michael Polanyi, *Personal Knowledge*, London, 1958, pp. 12–13; J. L. Synge, *Relativity, The Special Theory*, Amsterdam, 1958, pp. 161–2; and L. S. Swenson, *The Etherial Ether*, Texas, 1972.

were invoked to explain the discrepancy between the theory and the experiment, and it was the experiment, not the theory, that was disbelieved. The theory was retained because it was so rational, so profound, and had such great unifying power. As Tolman puts it "the results are to be accepted not only on the basis of the experimental verification which they have received in those cases where it has been possible to test differences between the predictions of relativistic and Newtonian mechanics, but also on the basis of the inner logicality of the theory which has led to them and the harmony of this theory with the rest of physics. The achievement of this logicality and harmony depends on the reconciliation of so many factors that we can feel considerable confidence in accepting results of the theory when necessary prior to their experimental verification."[3]

Radical empiricism accords too little weight to reason. It is largely because reason has been too narrowly construed. Reason has been taken to be purely deductive reason. A deductive argument is one where it would be inconsistent to allow the premisses and to refuse to concede the conclusion, just as an analytic proposition is one where it would be inconsistent to deny it. Such arguments are, indeed, very, very strong—because communication breaks down if people fail to acknowledge them—but very thin—they are purely verbal. The radical empiricists are quite right to point out that this sort of reasoning, useful and important though it is, cannot carry us all the way in natural science, but wrong to suppose that this is the only sort of reasoning. At the very least, despite, as I shall argue, Sir Karl Popper's claim to the contrary, some inductive arguments must be admitted; and, as I shall argue also, many other considerations, which are not simply a matter of experience, habit or prejudice, guide us both in our scientific theorizing and in our common-sense appreciation of ordinary life.

Popper[4] eliminates the problem of induction by claiming that the characteristic feature of scientific theories was not that they could be proved true by inductive arguments but that they could be falsified by experiment and observation. Falsifiability is for him the hall-mark of science. It is clear why this must be so. If we accept, as Popper does, that it is the aim of physics to discover universal laws of the

---

[3] Richard C. Tolman, *Relativity, Thermodynamics and Cosmology*, Oxford, 1950, para. 29, p. 53.

[4] K. R. Popper, *The Logic of Scientific Discovery*, English tr., London, 1959.

general form

$$(x)[A(x) \rightarrow Z(x)]$$

(where A may itself be complex), and if we demand, on empiricist principle, that laws of this form be logically related to particular observations, which must be of the form

A(i) & Z(i)   or   A(j) & $\sim$Z(j)
      or $\sim$A(k) & Z(k)   or $\sim$A(m) & $\sim$Z(m),

and if we demand, further, that the logical relation between them be deductive and not inductive, and that we are to argue from particular premisses to a general conclusion, then the only possible logical relation that will satisfy these requirements is that of inconsistency, which holds between

$$(x)[A(x) \rightarrow Z(x)] \quad \text{and} \quad A(j) \& \sim Z(j),$$

and which enables us to infer the falsity of the former from the truth of the latter. Thus if we are to fulfil Popper's requirements we shall be able only to falsify, and not to verify, suggested universal laws of nature. This is unsatisfactory. We do not believe that the laws of physics have merely failed to be falsified. We believe that they are, to some extent if not conclusively, confirmed by the accumulated evidence in their favour. In his later work Popper has attempted to give some account of the way in which scientific laws are confirmed or corroborated by empirical evidence, which brings his philosophy of science closer to the actual practice of scientists, although at the cost of blurring the stark simplicity of its original outline. Apart from this, Popper's account, although valuable, was misleading in that it did not explain why we should seek universal laws of nature or how we should select putative laws to put to the test of attempted falsification. It cannot be just an arbitrary preference on the part of physicists for formulating propositions of the form

$$(x)[A(x) \rightarrow Z(x)],$$

or they could equally well look for existentially quantified propositions such as $(\exists y)[A(y) \& Z(y)]$ or any other propositions of the first- or higher-order predicate calculus. Popper is, in spite of his diatribes against Plato, still a covert Platonist in his assumption that the fundamental truths about physical reality are encapsulated in

universal form. Moreover, he assumes that we are somehow pre-
sented with likely-looking candidates. But that will not be secured
to us merely by stipulation of logic. Many, many implausible laws
could be expressed in the form

$$(x)[A(x) \rightarrow Z(x)];$$

it is no good simply comparing scientific hypotheses with biological
organisms, and talking, as Popper does, of the survival of the fittest.
Falsification provides a quick death for unviable hypotheses but,
Popper's denials notwithstanding, needs to be supplemented with
some account of their birth and generation. Popper tries to produce
criteria of simplicity and strength whereby all possible hypotheses
can be ranked in order of inherent plausibility, so that instead of
choosing them at random to see whether they can be falsified, a
scientist can work through them systematically. This is correct, so
far as it goes. Not all hypotheses are equally strong candidates
initially. We have various grounds, of intellectual elegance, economy,
and intuitive plausibility, which enable a scientist to pick out likely-
looking laws for first consideration. But the grounds of selection are
wider, as well as being more difficult to formulate, than Popper
supposes. They are in fact witnesses to another strand in physics
than a purely empiricist one. For the present it is enough to note
that without some such supplement, the simple Popperian account
fails to do justice to the actual structure of physics and activities of
physicists. To put it crudely, if Popper's original account were true,
the Special Theory of Relativity would have been rejected when
Miller obtained a positive result on repeating the Michelson–Morley
experiment. But it was not rejected. Therefore Popper's account is
not true, being itself falsified by the actual course of scientific
theorizing.[5]

Although Popper's account does not succeed in eliminating the
problem of induction, it is none the less extremely valuable. Using
the distinction between NATURAL LAW and BOUNDARY CONDITIONS,
to which we shall return later,[6] it insists that our experimental observations, which discover the latter, are "theory-laden". We do not
just have experiences: we make observations, often in order to put

[5] Popper argues (*Logic of Scientific Discovery*, p. 46n.) that Miller's results were
discounted because they were not reproducible. But at the time they were very well
supported. It was thought they *must* be wrong, not because other physicists failed to
reproduce them, but because they ran counter to accepted theory.
[6] Ch. I, pp. 10–11, Ch. VIII, pp. 128, 140–1, Ch. X, pp. 159–62.

some question to nature, and see whether such and such a hypothesis is borne out. Many radical empiricists have failed to recognise this. They have felt that sense experience is some sort of basic given, the ultimate reality out of which all our knowledge must be constructed. It would take too long to give adequate consideration to the general philosophical thesis of phenomenalism which has had a pervasive influence in modern thought. Suffice it to say that it does run very much counter to our ordinary experience. We seldom have raw sense data except when suffering from concussion or fever. Our normal experience is not one of experiencing coloured patches in our visual fields, but of seeing trees, houses, and people. Our eyes and ears are trained and skilful in discerning what there is to be seen and heard, bringing to bear on the task many preconceptions about what there is to be seen and heard. The radiologist can read an X-ray photograph, where we can see only a blur, and we similarly *read* the world around us. The various visual and auditory *stimuli* we receive act as cues, and guide us into seeing and hearing, and saying—both to others and to ouselves—more than what was strictly given to our sense organs. Ordinary sense experience is not a bare collection of sense data, but is already organized and interpreted even as it is perceived, and is thus already more than a merely minimal report of the basic raw stuff of knowledge. Radical empiricism, therefore, cannot be argued for on phenomenalist grounds. Natural science is, undoubtedly, concerned with what is real, and is undoubtedly based on empirical evidence. But sense experience is not the only thing that is real, and is itself not simply something given, but largely also something sought.

If ordinary sense experience is already organized and interpreted, a scientific experiment is even more so. Scientific experiments, we believe, should be repeatable and explicable. Consider the reaction of scientists to so-called "miracles". Consider first the unique anomaly. It is at least conceivable that on one occasion an event should happen which was contrary to the laws of nature granted all normal conditions. Suppose the Archbishop of Canterbury were to levitate in full view of the television cameras at the State Opening of Parliament. Suppose, further, that a searching investigation was immediately undertaken by Fellows of the Royal Society, and that nothing unusual had been found. In that case we might say that it was an anomaly, a miracle, or an unexplained phenomenon, according to choice, but it would not have any great bearing on the normal scientific enterprise, just because it had happened once only, and science is concerned with repeatable observations. Suppose, secondly, that the Archbishop did it again, and appeared to be able to levitate at will. In this case his levitations would be in a sense repeatable, but still fall short of scientific repeatability, because they could not be repeated by any other competent person. It is part of the scientific ideal that science should be the same not only every when and every where, but for every person: a difference of person *per se*, should make no more difference than a bare difference of time or place. Archiepiscopal charisma is an unscientific property. Suppose therefore, thirdly, that many people could, after due meditation, levitate. Levitation would be as repeatable a phenomenon as sprinting. Not everyone can do it, but, given practice and natural aptitude, many a man can. But still the scientist would be unhappy. However repeatable such phenomena were, they would not be explicable according to the normal scientific canons. It is instructive to see why. It is partly that we have no theory which looks like linking a person's desire to levitate with the generation of a sufficient force to counteract the effect of gravity. But it is not only that we do not, as a matter of fact, have such a theory, but rather that we do not think we could have one. Unknown forces have been discovered in physics before now. Magnetic or electrostatic repulsion could give rise to phenomena akin to levitation, and physics has had to be radically recast to accommodate electromagnetism. At various times in its history practitioners of physics have been fairly sure what a physical explanation should look like, and in the last century no theory of electromagnetism was felt to be acceptable, unless it gave a fundamentally mechanical account of electromagnetic interaction in terms of elastic particles or fluids. We ought to be more humble now. But with all due humility, physicists would still find it hard to stomach levitation or the psi phenomena of the parapsychologists. This is because it is linking mental with physical phenomena. Since the time of Descartes, science has extruded the mental from the realm of science. Science is about *res extensa*, which is entirely separate from *res cogitans*. The sort of factors which can enter into a scientific explanation can be measured—that is the sole capability required of *res extensa*—but not experienced from inside—not cogitated. The idea of being able to suspend or counteract the laws of nature merely by willing it, however intelligible to the imagination, is entirely alien to the spirit of science. The whole of science has been marked by the exorcism of animism. Primitive man may have addressed spells to his trees and crops, because he ascribed to them a capacity to understand what he said and to obey it. Science is quite different. Although it could be the case that spells and incantations were effective in bringing about one's desires in nature, as they are, sometimes, in human affairs, nevertheless if this were, indeed, the case, it would break down the fundamental category distinction between persons and things. Hence the very deep-seated hostility, on the part of scientists, to parapsychology in general, and to miracles in particular. Although the scientist cannot rule out *a priori* the possibility of such things happening, he discounts it as much as possible, because it runs counter to the whole ideology of science.

The ideology of science can be more clearly appreciated if we contrast it with the ideals that inform other disciplines—history, law, philosophy, theology, geography, and, in a very different vein, pure mathematics and formal logic. These also are organized bodies of human knowledge, and in the Middle Ages would have been called *scientiae*, but are unlike the sorts of knowledge scientists seek, not only in the kind of evidence admitted but in the type of argument adduced and in the understanding looked for. Even among the sciences properly so called, there is a diversity of methods and aims. Physics, from the Greek word φύσις, meaning nature, is often taken to be the paradigm science to which chemistry, biology and all the other sciences should one day be reduced. There are objections to this view, which are not merely due to the *amour propre* of those being dubbed second-class scientists. Apart from the immense difficulties in actually carrying out the reductionist programme, it seems that biochemists, ethologists, geologists and the like are asking different questions from those the physicists are trying to answer. In some cases they are limited in space—biology and geology are effectively limited to the surface of this planet—in others they are explicitly temporal—geology, again, is concerned with the actual history of this planet, cosmology with the history of this universe. Physics, by contrast, is concerned with what happens every where and every when—not only what actually happens, but what would happen were circumstances to be different. Using the distinction between natural laws and boundary conditions,[7] we can characterize physics as always being primarily concerned with the former, whereas other sciences put more weight on the latter, either exploring

---

[7] K. R. Popper, *Logic of Scientific Discovery*, Ch. III, pp. 59 ff.

in greater detail what holds good subject to certain pervasive and important conditions—e.g. chemistry—or how, against the given background of general natural laws, special conditions yield particularly significant results—e.g. physiology, genetics. If physics concentrates on the formulations of fundamental laws of nature, it will be a "restricted science",[8] concerning itself only with very general phenomena, but not with the complexity of what actually happens. The physicist will tell us that if we consider a living organism together with its environment the total mass throughout the metabolic process will stay the same, and likewise the energy. But this is, to the biologist, profoundly uninformative information. The physicist is giving him, with great show of confidence, answers to questions he did not ask. We may be less unsympathetic than the biologist, and see that the conservation laws are conditions which have an important bearing on metabolism. But we have to concede that the physicist obtains his high degree of generality at the cost of not addressing himself to many questions we might naturally ask. Physics has stringent standards of irrelevance, and can operate only by excluding many topics and many features from its purview. We also must concede that often the force of an explanation lies in its elucidation of boundary conditions. Although it is open to the physicist to concern himself, on methodological principle, only with universal laws of nature, it is not open to him to deny by *fiat* the fact that many explanations are explanatory by reason of unravelling the complicated interplay of particular factors. To explain in detail how an organism obtains its energy furthers our understanding of the organism whereas merely to chant the First Law of Thermodynamics over it does not. Explanation often involves a lot of jig-saw work, and to pass this off as a "mere special case" is to miss the whole difficulty of the discipline.

Nevertheless, physics is an entirely legitimate part of science, and one that illustrates an important part of the scientific ideal. In its concern for universality and rationality, and in its unconcern for the particular merely as such, physics is very Platonist. It is as Platonist as it can be, while still remaining empirical. The ideal of the scientific experiment, which should yield observations that can be repeated by anyone, any where and any when, is fashioned in accordance with

---

[8] See C. F. A. Pantin, "The Ballard Mathews Lecture" in *Science and Education*, Cardiff, 1963, p. 12; quoted in C. F. A. Pantin, *The Relations between the Sciences*, Cambridge, 1968, p. 173; see also ibid., Ch. I, esp. p. 18.

Plato's prescription that the only real truth is universal truth, time-less, placeless and impersonal; or, as the Early Church maintained, it is a mark of a true proposition of religion that it could be recognised *semper, ubique et ab omnibus*. Only, whereas Plato thought that knowledge of nature could be obtained by pure reason alone, the modern physicist has made some concession, albeit a minimum con-cession and in as Platonic a form as possible, to empiricism. Instead of radical empiricism, we should espouse some slightly chastened form of rationalism, if we want to follow the ideals of modern physics.

The principles of natural philosophy are, therefore, much less simple and much less coherent than is suggested either by the de-ductive rationalism of Plato or the radical empiricism of modern times. There is neither a single starting point—the ἀνυπόθετος ἀρχή, unpostulated principle, of Plato or the sense data of the pheno-menalists—nor a uniform method. Worse, our different intimations of reality may work in opposite directions: as rationalists we may seek rational transparency in the laws of nature while as empiricists we acknowledge we must experience some opaque contingency in our encounters with the external world. We seek simple, coherent theories, which explain why things are as they are, but at the same time we expect facts to be brute, and believe that only empiricism in some form or other will subject us to their brutal discipline. Besides our intimations that the real is both rational and yet recalcitrant to our will, which in turn give rise to further intimations of objectivity, universality and necessity, we have vague ideas of substantiality—what entities really exist—and of space, time and causality. Many of these have started by being opinions of common sense, which have then been refined and clarified as men have sought to be more exact and explicit in their understanding of nature. Contrary to what the radical empiricists maintain, there is no sharp distinction between observation statements and theoretical statements, but rather a gra-dation; and the interplay of different principles is much more com-plicated than they suppose, but open to correction by empirical evidence, contrary to the opinion of Plato and the deductive ra-tionalists. For the most part we shall be concerned with elucidating the various rational principles on which we attempt to reach an understanding of Nature, as distinct on the one hand from our mathematical thinking which is indeed purely *a priori*, and on the other from our understanding of people and from the humanities generally which demand gifts of sympathy and insight far more elusive and difficult for the philosopher to formulate. Within this range we shall start with the simplest canons of inductive inference, without which it would be impossible to have any understanding of the world around us, and then we shall extend and generalise this so as to exhibit the rationale of our concept of causality and the presuppositions about time and space which underlie many scientific theories. But we need always to be reminding ourselves that, however rational, coherent and elegant a scientific theory is, it does not prove absolutely that it is true. It always could turn out to be false, because the world always could (logically could) turn out to be different. The most we can do is to show how rational the theories which in fact do account for the phenomena are, and the most that we can say for them *a priori* is that if the world was different and they were false, then the world would be less rational than it might have been. We cannot, as Plato sought to do, show that the world must be of such and such a form: but often, reviewing theories *ex post facto* we can discern an underlying rationality within them, and with the benefit of hindsight may persuade ourselves that the world had to be basically of the form that it is, *if* it was to be the most rational of all possible worlds.

Further Reading

B J. C. Graves, *The Conceptual Foundations of Contemporary Relativity Theory*, Cambridge, Mass., 1971, §§ 2–3, pp. 7–42.

A   Bertrand Russell, *Problems of Philosophy*, Oxford, 1946, Ch. VI, pp. 60–9. OPUS edn., 1967, pp. 33–8; reprinted in Richard Swinburne, ed., *The Justification of Induction*, Oxford, 1974, Ch. I, pp. 19–25.

II

Induction

THE simplest arguments about the natural world involve some principle of induction. For example, we may start from the fact that all the swans we have ever seen have been white, and infer that all swans are white. Or we note that the sun has risen once in every twenty-four hour period from midnight to midnight, as measured by a caesium clock,[1] and on the basis of this predict that the sun will rise tomorrow too. Or we observe that people to whom arsenic has been administered have subsequently died, and conclude that arsenic causes death. These inductive arguments take many different forms: they may argue from a certain number of particular instances to a generalisation, or they may argue from a certain number of particular instances which we have observed to another particular instance which we have not observed. They may argue from past events to future events, or from known past ones to unknown past ones. They may seek to establish invariable concomitances—the whiteness of swans—or they may seek to establish some causal relation.

The different forms of inductive argument are interconnected. Thus if we can legitimately argue that all swans are white, then if we know of any particular bird that it is a swan, we can argue purely deductively that it is white. And *per contra* if for *any* day we can legitimately infer that the sun will rise, then we have established that it will rise *every* day. For various reasons, which will become apparent later, I shall concentrate on the former type of inductive argument, where we argue from particular to general, rather than

[1] The familiar example of the sun's rising tomorrow can be misleading. If we do not take care, it will turn out to be a tautology, for days are often defined in terms of the rising and setting of the sun. The time measured by a caesium clock does not depend upon the rotation of the earth, and therefore we can specify our common expectation of the sun's rising after a certain period of time in terms of this.

the latter where we argue from particulars to a further particular; and in the same vein I shall start with the general uniformity of nature rather than some law of causation, and with being able to make omnitemporal claims on the basis of temporal observations, rather than simply predicting future events on the basis of past ones. But other approaches are equally possible, and if any one form of inductive argument is legitimated, the others are established too.

The argument about swans could be put more formally thus:

All the swans I have ever seen were white.

∴ All swans are white.

The premiss could be expanded thus:

This bird is a swan and is white.
That bird was a swan and was white.
A third bird there is a swan and is white.

. . .

A 257th bird there was a swan and was white.
I do not know of a swan which was not white.

∴ All swans are white.

The fuller version has the advantage of making the importance of the last premiss more apparent.

As we have stated, this argument is not a deductive argument; that is, it is not self-contradictory to assert the premisses and to deny the conclusion. A man would not be contradicting himself if he asserted that all the swans he had ever seen had been white, but not all swans were white. Indeed, he would not only be not contradicting himself, he would actually be telling the truth. Not all swans are white. Some, in Australia, are black. They provide an admirable proof of the distinguishing feature of inductive arguments, that they are not watertight. It is possible using an inductive inference to start from true premisses and end with a false conclusion. Nothing venture, nothing win. Inductive arguments are risky, but that is because they offer us something worth having, which we would not have without them, namely knowledge we did not have before.

It follows that if we seek a justification of induction, we cannot hope to get a deductive justification, for that would end by showing induction to be, what we know by definition it is not, deduction. I need to emphasize this point, because it is always easy in philosophy to make an argument watertight by redefining some of the terms so that it becomes deductive; and many philosophers have done just this with induction. One could say that a swan is not a swan unless it is white, or to take a slightly more plausible example an acorn is not an acorn unless it will grow into an oak. Often, particularly in natural science, we do redefine substances so that what had hitherto been contingent features become defining characteristics. If was first a synthetic truth that sodium salts would give a flame a characteristic orange hue of two very closely-linked wavelengths: but now we should hesitate to call anything sodium which did not show these "D-lines" in the spectroscope. Especially in chemistry, we have often saved the truth of some synthetic proposition by redefining our concepts or introducing new distinctions so as to make it analytic. Thus we introduce the concept of allotrope to accommodate the fact that graphite and diamond, though both the same chemical element, carbon, do not have the same properties: or that α-sulphur, β-sulphur and amorphous sulphur melt at different temperatures, have different shaped crystals and slightly different colours; or that white phosphorus is highly inflammable where as red phosphorus is not. Again, at a later stage, chemists introduced the notion of isotope, rather than completely abandon the idea that atomic weight was a fundamental property of atoms; that is, instead of the synthetic truths that the atomic weight of hydrogen is 1.008, of oxygen 16.00, of uranium 238.07, we have the analytic ones, that the weight of protium is 1, of deuterium 2, of tritium 3, of ^{16}O 16, of ^{17}O 17, of ^{235}U 235, of ^{238}U 238, etc. Thus it is following in the best scientific tradition to deal with counter-examples by redefining some terms so that the counter-example is no longer a counter-example. Nor do we know in advance which propositions we shall save in this manner. And therefore it is very easy for a philosopher, concerned to secure inductive argument from the possibility of refutation, to respond to counter-examples in like manner. Whenever one appears, he strengthens the threatened truth by making it analytic. And since inductive inference, though liable to error, is not normally erroneous, the philosopher is not driven to do this so often as to be palpably fraudulent. We need, therefore, to be especially careful, when considering the justification of induction, not to fall into this trap.

Inductive arguments are not deductive, and therefore no deductive justification of induction is possible. So much may be conceded, and it may be asked whether there is anything more to be worried about. It has been suggested that no justification of induction is possible or necessary. We make inductive inferences, and we must just accept that we do. We can, if we like, draw up canons of inductive inference by which to select good inferences from bad ones, but we cannot try and justify inductive inference as such. We can justify particular inferences by showing them to be good inductive inferences according to the accepted canons, but we cannot justify inductive inference generally. To ask whether inductive inference, as a species of inference, is valid is like asking whether the law is legal.[2] The suggestion has much to commend it. It deals with the sceptic's doubts in the most effective possible way, by ignoring them. Unless and until the sceptic can give some grounds for his doubts, the philosopher will not take them seriously. After all, although it would not be self-contradictory to affirm the premisses and deny the conclusion of an inductive argument it would be unreasonable to do so. The only grounds on which I can reasonably deny that all swans are white are that I know of black swans. If I do know this, then one of the premisses is not true: if I do not, I have no reason for denying the conclusion, and I am contravening the rules of discourse among reasonable men if I idly query what other men assert without having any justification for my queries. So much may be said for the view that induction needs no justification, and for many people it is enough. And yet a residual discomfort remains. The short way with scepticism is too short. After all, many philosophers have raised the question of the validity of inductive inference, and to dismiss their questions out of hand is cavalier. It is better to answer questions than not to let them be asked. And we may wish to explain to ourselves, without ever doubting the validity of inductive inference, the way in which it acquires its validity. We can give ourselves an account of why deductive arguments are valid—they are a form of rule-observance, and if we fail to observe the rules of language, communication breaks down, and we shall find ourselves unable to convey our meaning to our hearers: and we should like to give ourselves some similar account of the validity of induction. The question does not seem as silly as that of why the law is legal. It is rather like the question of why, and to what extent, legal obligations are moral ones. We want to know to what extent we ought to accept inductive inferences, and why.

---

[2] P. F. Strawson, *Introduction to Logical Theory*, London, 1953, p. 257.

Before we embark on the task of justifying induction, we must raise the question 'Justification to whom?'. To justify induction is to justify it to some sort of questioner who has some sort of objection to, or doubt about, induction. There are several distinct objections and doubts that can be levelled against inductive inference, and as the difficulties vary so also must their justification. A Platonist should find little difficulty in induction. He believes that universals are real; indeed, only they are really real. The significant truths are those that deal with the relation of universals, types, sorts or kinds. A true statement will be of the form **A**-ness involves **B**-ness, or **A**-ness and **B**-ness are incompatible. We, poor mortals, being unable to see **A**-ness or **B**-ness in themselves, but only instantiated in particular objects, events, or cases, cannot ever know for certain the exact relations that subsist between **A**-ness and **B**-ness. We can only hazard guesses about what the relations might be on the inadequate evidence available to our inadequate eyes. Our guesses always may be wrong. But at least we know the form they must take if they are to be right. And if we allow that the world we can see, although an imperfect, is not altogether a misleading, exemplification of the ultimate reality, then we can go further. *Ab esse valet consequentia posse.* We can by finding instances of **A**-ness which are also instances of **B**-ness prove that **A**-ness and **B**-ness are compatible, and so falsify the hypothesis of their being incompatible. We can operate the procedure of eliminative induction in the usual manner. Although we cannot reach deductive certainty unless we have some further postulate limiting the number of possible universals or types of which a particular case could be an instance, we have the assurance we looked for.

The assurance of the Platonist is not available to many : but often, though they are not prepared to believe in the existence of Platonic universals, they are prepared to believe in causality, or in the existence of natural laws, or of uniformities or regularities in nature; and this will give them the assurance looked for, although it is far from guaranteeing that on any particular occasion we have discovered a causal relation or a natural law, or that what we take to be a causal relation or a natural law will continue to hold. But then we never set out to produce a watertight guarantee for our inferences, but only to give *some* reason for accepting inductive inferences, and this the doctrine of causality or the existence of natural laws does do. It is the same with universals: if once we are persuaded that a certain type of thing exists, whether universal, or law, or cause, we are prepared to recognise a particular specimen on less than foolproof evidence. A chemist, determining the melting point of a crystalline substance, makes only a few observations, many fewer than a philosopher observing swans or the rising of the sun. Just as with material objects we discount the possibility of mistake, illusion, or hallucination, and are prepared to make statements about some material objects on less than the best possible evidence, in spite of the fact that we might subsequently have to eat our words, so, if we believe that laws do exist, whatever we understand by this, we mean at least that we are prepared to recognise events as instances of natural laws on less than perfect evidence, even though we may sometimes turn out to have been wrong. A phenomenalist, who does not believe in the existence of material objects, experiences nothing but his sense experience when he sees or hears or feels a thing: whereas the ordinary man, who does believe that material objects exist, does not merely have sense data, but reads into his sense experience more than he actually experiences, and sees or hears or feels *things*. In one important sense the phenomenalist and the ordinary man have similar experiences; neither has any sense-datum that the other does not have. Yet while the phenomenalist is merely experiencing what he is experiencing, the ordinary man, because he believes that there are material objects which can be seen, heard and felt, is able to see, hear and feel, material objects. Of course, he is sometimes wrong: he may see as an elephant what subsequently turns out to be a haystack, whereas the phenomenalist who confines himself to describing his visual experience without engaging in any ontological commitments, is absolutely safe. This security, most of us feel, is bought at too high a price. Believing in the existence of the external world, we are content to stick our necks out, and run the risk of being wrong, by seeing things, and talking about them, as material objects, which are relatively understandable, rather than dumbly observing the kaleidoscope that happens before our eyes.

The analogy between the existence of natural laws and the existence of material objects is widely felt; but it is difficult to articulate clearly, and many philosophers shy away from it as a muddle-headed excursion into metaphysics. Nevertheless, strongly held convictions should not be given a brusque dismissal, even though we cannot immediately give an adequately clear account of what is being maintained and on what grounds. The notion of natural law is itself imprecise. It does not cover every conceivable correlation—else to say that there were natural laws would be to say nothing, because every possible course of events would be an instance of some conceivable natural law. We rule out laws formulated in terms of phoney predicates, such as Nelson Goodman's "Grue" and "Bleen",[3] and very complicated and ugly functions; but we cannot lay down in advance what laws we will and what we will not accept. We might start by saying that we prefer linear functions to quadratics, and, generally, polynomials of lower degree to those of higher degree; but often we prefer sine or exponential functions most of all. The simplicity and rationality that a modern physicist sees is not always immediately obvious, and sometimes was not readily accepted by scientists themselves. Thus the formula

$$\frac{\partial^2 V}{\partial x^2} + \frac{\partial^2 V}{\partial y^2} + \frac{\partial^2 V}{\partial z^2} - \frac{1}{c^2}\frac{\partial^2 V}{\partial t^2}$$

seems very elegant and simple, though it would need a long course in physics to explain why it was elegant and simple; and the formulae of the Special Theory of Relativity and the whole calculus of tensors seemed very clumsy at first. It seems that we cannot give an exact account of what we mean by natural law or simple regularity because we do not altogether know yet. We are tutored by experience. In our efforts to understand nature we find theories forced on us by the experimental facts, which are not particularly acceptable at first; it is only later we come to appreciate their rationality. Our notions of elegance and simplicity are not completely determined in advance, but are developed and refined in the attempt to render experience rational.

The content of our belief that natural laws exist is therefore necessarily not completely clear: the grounds for the belief can, however, be clarified. Natural laws are thought to have an existence analogous to that of material objects, because like material objects they have a certain non-subjectivity, and a certain ineluctability. The external world is, by the very definition of the concept, common to all observers: and in the same way, natural laws if they are to count as natural laws must be discoverable by all competent scientific observers. Indeed, in this respect, a natural law is very much more real than a material object; for a material object exists only in a particular place over a particular period of time, whereas a natural law operates

[3] Nelson Goodman, *Fact, Fiction and Forecast*, London, 1955, pp. 74 ff.

everywhere and always. Natural laws are in this like Platonic universals: though not themselves visible or audible or tangible, they are more real than objects of sense, because universal and not limited by the particular conditions of space and time that necessarily limit each material object. The notion of reality is in part connected with that of rationality, and that of rationality with that of universality. The uniformities of nature, if they can but be found, are candidates for high-grade ontological status which may well be formulated by saying that they exist.

Dr Johnson's refutation of phenomenalism was to kick a stone. In doing this, he was playing on the other facet we noted in our conception of a real thing, that real things are potential obstacles to our will.[4] The rainbow can be seen by everyone, but is not a real thing because no one has ever collided with it. Impenetrability marks out our concept of a real thing. It is, as we saw, the opposite of rationality. Things exist whether we will them or no, whether we can understand them or cannot. They are the brute facts of our lives, which we may try to comprehend if we please, but whether we succeed or fail, must accept all the same. The same is true of the laws of nature. Just as we cannot walk through a brick wall, so we cannot without getting burnt put our hands in the fire. We may not understand how a particular poison works, but if we take it we shall surely die just the same. The law of gravity is as much part of the furniture of our lives as the ground on which we stand. The laws of nature are ineluctable.

The notion of ineluctability is that of impossibility: and the notion of impossibility is set against that of endeavour. It is when we discover that no matter how we try, a desired result does not follow, that we conclude that it cannot be obtained. Hence the crucial importance of experiment. Trial and failure is our method of exploring the impossibilities which surround us. Experiment is, so to speak, a probe with which we feel out the four-dimensional solidities which stand in the way of our achieving our ends. Once we have discovered what we can and cannot do, we have equally discovered what things are possible and impossible, and what things are necessary, and so have obtained the concept of a causal relation or a natural law.[5]

The exact logic of the foregoing arguments has not yet been made clear and we shall need to elucidate the argument from experiment

[4] Ch. I, p. 12.
[5] See, more fully, Ch. III, pp. 35–6.

more fully in the next two chapters. In any case, a determined sceptic could still refuse to be persuaded to make the metaphysical commitment that natural laws exist. We have given the arguments in their present rough form, because it is in that form that many non-sceptics hold them. For nice-minded philosophers, who disdain all metaphysics and statements of ontological privilege, we can transpose the argument from rationality into a more purely logical form. It is taken as obvious—perhaps a last vestige of Platonism—that the best sort of sentences are those in an impersonal third person and in a tenseless present tense, for example 'Snow is white' and 'Two and two make four'. Sentences of this sort mean the same in whosesoever mouth they are spoken, whensoever they are spoken, and wheresoever they are spoken. Only these sentences express pukka propositions. Other sentences do not express propositions because they contain, explicitly or implicitly, "token-reflexive" or "indexical" terms. By token-reflexive, or indexical, terms are meant words such as 'I', 'this', 'now', 'he', 'here', 'there', 'then', words whose reference depends on the time and place of utterance or on the person who utters them: 'I' in my mouth means me, in yours you; two people talking over the telephone can say the one 'It is raining here', the other 'It is not here' without contradicting each other, because the word 'here' is being used to refer to different places. Differences of place, of person, or of time are, we assume, in themselves irrelevant to scientific truth.[6] Tenses are implicitly token-reflexive: at different times it is appropriate to say 'Hitler will unleash a world war', 'Hitler is unleashing a world war' and 'Hitler unleashed a world war'; and in the same way the words 'future', 'present' and 'past' are token-reflexive. Therefore the sentence 'The future will be like the past' does not express a proposition at all, and the question that could have this for an answer is not a proper question. Similarly if we describe induction as inference from what we know to what we do not know, the word 'we' introduces an illegitimate token-reflexive element. Once we ban token-reflexive terms altogether the problem of induction cannot be posed, and so we need not be worried by it.[7]

The guiding principle of this argument is non-egocentricity. The distinction between future and past, being a projection of one's own temporal position, is irrelevant. The particular instances I happen to know of are not thereby distinguished from those I happen not

---

[6] See, more fully, Ch. IV, pp. 57, 59–60, Ch. V, pp. 70–2.
[7] A. J. Ayer, *Philosophical Essays*, London, 1954, pp. 188–8.

to know of. My particular observations are typical of the general run, because the fact of my observing them is completely unimportant, and therefore the ones I happen to observe are bound to be a random sample. The problem of induction, it is suggested, is like other problems raised by sceptics, not a real problem at all, but only the symptom of the philosopher's neurotic obsession with his own self. The sceptic about induction keeps on asking 'How can I know about what I have not myself observed?' just as the phenomenalist and the solipsist do. To such a question no answer is necessary for the man who does not regard himself as absolutely a special case, and no answer is possible for the man who does. Scepticism is thus seen as a sort of pride, the epistemological form of original sin. What the sceptic needs is not an assurance about the nature of the world, but the grace of humility, so that he should be ready to accept arguments, and should not demand always an armour-plated guarantee that he may never be found to be in the wrong. If he is to be reasonable at all and enter into the universe of rational discourse, he must abandon his obsessive self-centredness, and learn to view himself as only one person among others, and not to be distinguished from them in any fundamental way. Rationality is opposed to egocentricity, and to the extent a man is being rational he must come out of himself and stick his neck out: and *per contra*, to the extent that a man insulates himself in the fortress of his own self-sufficiency, he isolates himself from everything which is not himself, and cuts himself off from all understanding. To be invulnerable is to be invincibly ignorant.

That acceptance of inductive arguments is a necessary condition of the possibility of discourse can be argued for in less theological terms. It is clear that language presupposes some sort of uniformity in the world. We could not apply the words 'table', 'man' or 'tree', unless there were stable concomitances of features to which we could apply these words. If our experience were kaleidoscopic, we could not talk about it to anyone else, because we could not get him to identify and re-identify what we were talking about. Language is a matter of observing linguistic rules, and language can be applied to experience only if experience is also in some way regular. Moreover, not only could we not talk about the external world if it was not fairly regular, but we could not learn to talk at all. If the universe dissolved into chaos but we continued as disembodied spirits, we might be able to communicate by some mysterious means with one another, but our topics of conversation would be severely restricted and we could never admit anybody new to our circle. The possibility envisaged by the sceptic about induction is conceivable, though barely so: but the fact that he continues to use language, and goes on talking to us about the course of future events, shows that he does not practise his doubts about the future.

The argument from the possibility of discourse is often regarded as the modern version of Kant's argument for causality from the possibility of experience. In part it is: many of Kant's somewhat psychological arguments about the nature of thought can be exhibited more satisfactorily as logical arguments about the nature of language. Nevertheless, in this case, the simple psychological argument does also hold and is epistemologically relevant. A man suffering from concussion may still have sense data. He hears sounds, and sees shapes and colours swirling before his eyes: but it can hardly be called experience; the man has no grasp of it, cannot describe it, not even to himself, cannot organize it into anything recognisable and coherent. To count as experience, the various sense data must be grouped together and perceived as definite wholes. The Gestalt psychologists have investigated the way in which the sense organs do organize sensations into comprehensible units. The reader who doubts the importance of this should take up the use of a microscope, and discover for himself how useful diagrams, and how useless photographs, are. The photograph must be the same as the image that falls on the retina, unorganized and incomprehensible: the diagram shows what the skilled observer sees, and shows us what to look for—how to see the specimen. Our everyday experience is similarly organized. We see tables, trees and men as tables, trees and men, and if we do not see them as such we can barely be said to see them at all. In order to have experience we need to be able to categorize it, and the inductive sceptic's suggestion of an uncategorized experience is not only inconceivable but, in view of the nature of experience, logically impossible. Hence, there must be uniformities in any world that can be experienced; and, in particular, those uniformities we express in the form of causal laws.

THE various justifications of induction I have outlined here seek to justify induction to different sorts of questioner. Different philosophers take different things for granted, and explanations of why inductive arguments are to be accepted will vary with what each philosopher is prepared to take as a starting point. Metaphysically-minded philosophers are prepared to discuss whether natural laws exist in the way that universals or material objects are said to exist, whereas philosophers in the modern fashion may find it illuminating to see inductive scepticism as a case of scepticism generally and may be led to accept the validity of inductive arguments by considering what else would have to be abandoned if induction were to go. In the end, each philosopher will want a different justification: not only will he find many of the lines of argument suggested here irrelevant, but if any seems to have some bearing on the matter, he will want to argue the matter very much more fully before he will be convinced. It would seem that we have at best provided a poor justification of induction, having said at the same time too much and too little to satisfy the questioner. It may be asked whether a more direct and specific justification of induction could not be offered, which could be addressed equally to all questioners, and would be quite explicit, and take nothing for granted. It is to meet this specification that the last justification, immediately following, is offered. It is a pragmatic justification, intended for any rational agent whatsoever, and purporting to show that the "inductive policy", the policy of accepting inductive inferences, is the most rational policy a rational agent can adopt.

The argument falls into two parts. First that it is rational to try and make inductive inferences of some sort, and secondly that inductive inferences as we have them, at present, are the most rational way of making the attempt.

It is rational to try and make inductive inferences. For if we try, we may succeed: if we do not try, we are certain to fail. To take the case of predictions: if I make predictions, they can at worst come out wrong: if I do not make them, they cannot even at best come out right. And I must want to be able to predict. Situated as I am, with the future ever coming upon me and always unknown, it cannot be a disadvantage to know it in advance, and it may prove a help. Indeed, if I am to act at all, I must be able to predict the outcome of my actions. I could not so much as take a step across the floor if I was not able and willing to predict that the floor would not give way. Therefore we must try to predict.

If we are trying to predict the future all we have to go upon is the past: or, to put it another way, if we are seeking to gain knowledge of what we do not already know, the only evidence we can have is what we do already know. The question is, how to use this evidence? Are we to assume that the future is like the past, the unknown like the known, or are we to assume that they are unlike? Will things happen in the same way as they have happened hitherto, or in some different way? Either alternative is logically possible, but only one can be a working assumption. For whereas the assumption that things will happen in *the same way* as they have done hitherto, enables us, in conjunction with the knowledge of how they have happened hitherto, to say how they will happen, the assumption that they will happen in *some different way* does not enable us to say anything at all. For whereas there is only one *same way*, there are many different *different ways*, and there is no reason to choose one rather than any other. To say a person is unlike somebody else is to say very much less about him than to say that he is like him: and similarly, to say the future will be different from the past is to say very little about it, too little to guide us in our predictions. The only assumption that will enable us to use the only evidence that we can have is an assumption of sameness. Only this is definite enough. Therefore, if we are to make predictions about the future at all, the only evidence we can have is the past, and the only assumption that we can use is that the future is like the past. And this is just what inductive arguments assume.

Further Reading

B   Richard Swinburne, ed., *The Justification of Induction*, Oxford, 1974.

The original proponent of the "Pragmatic" justification was Hans Reichenbach. See his

C   *Experience and Prediction*, Chicago, 1938, Ch. 5;

and his

C   *The Theory of Probability*, University of California Press, 1949, Ch. 11

See also:

B   Wesley C. Salmon, "The Pragmatic Justification of Induction", reprinted as Ch. V in Swinburne.

A David Hume, *An Enquiry Concerning Human Understanding*, Section VII; §§ 48–61, pp. 60–79.

III

Causality

CAUSALITY plays a large part in science. It can be elucidated in terms of laws of nature, themselves to be established by inductive arguments; but often philosophers have treated it separately, and have regarded cause, along with time and space, as key concepts for our understanding of natural phenomena. This was particularly true in the eighteenth century, when the philosophers who succeeded Descartes were attempting to elucidate our conceptual structure on Rationalist or Empiricist principles. Descartes made the question 'How do I know?' the fundamental one in philosophy. His method of Cartesian doubt led him to reject all the common opinions and assumptions he had inherited from earlier generations or shared with his own, and to accept as true no proposition until he had assured himself that he had indubitable grounds for believing it. He believed that by re-examining all his ideas in the light of this question, he could purge himself from all erroneous opinions, and be sure that he really knew the remaining propositions he thought he knew. One way, which proved very popular among philosophers of the seventeenth and eighteenth centuries, of posing the problem was to ask about the *origin* of ideas, 'Where do our ideas of space, time and causality come from?' But that confused two separate questions, one about the concept, the other about the application of that concept. Descartes himself assumed that if he could give a satisfactory account of the origin of our concepts, he had given a complete answer to the epistemological—'How do I know?'—question he had posed. He considered the concepts of which he could form "clear and distinct ideas", and regarded these as being secure and above reproach. Against this, Locke and the English Empiricists argued that the only source of knowledge is sense experience. But really they were answering a different question. Whether or not our concepts of space, time and causality are formed from sense experience, it is very plausible to maintain that it is only on the basis of sense experience that we are justified in applying those concepts to particular cases. It may, or may not, be the case that we have particular spatial, temporal or causal experiences from which we derive our concepts of space, time and causality: but at least we should doubt very much the reliability of a man who claimed that one thing was bigger than another, or that one thing happened before another, or that one thing was the cause of another, without basing any of these judgements on any sense experience whatsoever, his own or another's. It was only at the end of the eighteenth century that Kant made this distinction, and to that extent secured a reconciliation between the Rationalists and the Empiricists. The importance of the distinction is still often missed. The Logical Positivists in the 1930s were accustomed to maintain that the meaning of the sentence was its method of verification. But this seems to be the reverse of the truth. The method of verification may have an important bearing on the meaning, but essentially and characteristically does not exhaust it. We see this particularly clearly when we consider the meaning of, and the rules for the application of, the concept of cause.

Hume criticized causality from an empiricist standpoint. He believed that all our concepts were derived from sense experience (or, as he would have put it, all our ideas were derived from impressions). He asked what sense experience could possibly have given us the idea of causality, and discovered that there was no sense experience of causality that we could have got our idea of it from. He concluded that when we thought that one thing was a cause of another, this view was not derived by any form of reasoning from any sense experience, but was the result of habit. Causal reasoning was nothing more than a conditioned reflex. From having often encountered the one thing followed by the other, we got accustomed to expecting the latter whenever we saw the former; and this habit of mind expressed itself in a belief that the one thing was the *cause* of the other.

It is easy to find fault with Hume. He asked the wrong questions, and his theory of knowledge, as we said in Chapter I, was based on a "thin" view of reason, and a "flat" view of experience, and was therefore unable to account for our knowing many of the things we do know. His sceptical conclusions were, in consequence, mistaken.

But much of his critique of causality is, nevertheless, of permanent value. He has shown first that causal relations are not discernible by the senses, secondly that they are not particular relations, and thirdly that they are not analytic. There are important logical differences between saying that A is the cause of B, on the one hand, and saying that A is the same colour as B, or that A is to the right of B, or that A is bigger than B, or that A happened after B, or that A entails B, on the other. If a man reports that A is the same colour as B, I cannnot disagree with him without impugning either his veracity or his competence as an observer. But if a man says that A is the cause of B, I can disagree with him without imputing to him either dishonesty or incompetence. I can disagree with him, and put the matter to the test by repeating the experiment and seeing if I could get an A without its being followed by a B; and if I did, I should have established that A was not the cause of B, although it may well have been that the person who thought that A was a cause of B had invariably found A's to be followed by B's and so had very reasonable grounds for believing that A was the cause of B. Although we are entitled to apply the concept of cause only if we have adequate empirical evidence, the concept has built into it more than its rules of application. You can have adequate evidence for maintaining that A is the cause of B, but if you do maintain that A is the cause of B, then you are saying something which goes beyond the evidence you have at your disposal. You are saying not merely that whenever you have seen an A, it has been followed by a B, but that if ever there were another A, it would be followed by another B. And this latter is an open-ended claim towards the future, which is always vulnerable, and can never be established beyond all logical possibility of refutation. This shows both that the meaning of the concept of cause is more than its rules for applicability, and consequently that the causal relationship is not discernible by the senses, in the way that colour relationships are. Similar considerations show that the causal relationships are not particular. If I say that A is to the right of B, or that A is bigger than B, or that A is after B, the spatial or temporal relationship is a particular one between a particular A and a particular B, and could be different for *this* A and *this* B without there being any difference in the temporal or spatial relationships between any other A's and any other B's. But if I say that A is the cause of B, then I am speaking not only of *this* A and *this* B, but of *any* A sort of thing and *any* B sort of thing. Causal relationships are universal, or perhaps better, universalisable in a strong sense, that is to say, repeatable.

Hume has shown thirdly that causal connections are not analytic. He claimed to have shown that they are not necessary, but he used the word 'necessary' in a very limited sense which we should naturally render as 'logically necessary'. Certainly causal necessities are different from logical necessities. If B is a logically necessary consequence of A, then it would be self-contradictory to describe a situation as one in which A was present but B was absent. There are many such logical correlations in our language. I cannot be a parent unless I have a child; a thing cannot be red without its also being coloured; a married bachelor is a contradiction in terms. In all these cases our rules for the use of the words involved are such that we cannot deny the relevant proposition on pain of flouting some linguistic rule. We call such propositions analytic, and say that they are logically necessary truths. The word 'logical' is unfortunately equivocal, and can be used in several senses. The sense here used is that of deductive logic, in which we contrast the logical necessity of tautologies and analytic propositions with the physical, biological, social, or moral, necessities that we may encounter in other universes of discourse. On other occasions, however, we use the word 'logical' in a wider sense, roughly equivalent to 'rational', as when we contrast it with 'emotional', and say women are not logical, or some political speech is not logical. In other contexts we use it with some other shade of meaning; we sometimes talk of the logic of a situation; we sometimes contrast the logic of an argument with the facts on which it is based; we sometimes contrast logic with evaluation or interpretation. For these reasons philosophers have often been confused in their use of the word 'logical', and have maintained that causality was a logical concept, and have thought of causal connexions having some sort of logical necessity. Hume forces us to think more clearly. However necessary or inevitable it is that fire burns and water wets, it is not logically necessary in the sense that there would be no logical contradiction in describing a fire that did not burn a person or a thing which approached it, or some water that did not make things wet. Moses' report of the burning bush is not a contradiction in terms, and chemists can produce flames which are cold and do not burn either a hand placed in them or set alight ordinary combustible materials. Equally, although water does in fact wet, there are many other liquids, such as mercury, which do not, some of them extremely like water in appearance and behaviour. It would be perfectly possible to describe water in terms of its chemical

composition or other behaviour, and then there would be no contradiction in describing this water as not making anything wet.

Hume's point is a difficult one, because there are many apparent
counter-instances. Many causal regularities have been incorporated
into our language. The very words 'cause' and 'effect' are logically
connected, and there is a contradiction in describing a situation as
one in which the cause is not followed by its effect. Hume allows
that in this sense causes logically necessitate their effects, but argues
that this description begs the question. In the same way, as we saw
in Chapter II, it is part of the meaning of the word 'acorn' that an
acorn, given suitable conditions, should grow into an oak; and it is
part of the meaning of the word 'food' that it should feed one, and
of drink that one should be able to drink it, and of poison that it
should produce untoward consequences in anyone to whom it is
administered. Nevertheless, although it is part of the meaning of
these words that the things described should have the specified consequences, there are other more neutral descriptions available, which
do not carry causal connotations as a matter of logical necessity.
And it is in terms of these more neutral descriptions that the problem
of causality ought to be posed, and then no logical necessity can be
imputed to causal relations.

We may summarise Hume's negative critique as showing that the
causal relation is

(1) not given by the senses,
(2) not particular,
(3) not deductive.

These are valuable insights. His positive account, however, is unsatisfactory. He attempts to elucidate the concept of cause almost
entirely in terms of repeatability, or constant conjunction as he terms
it, partly because of his general view of the working of the human
mind, partly because of his confusion between the meaning of a
word and the rules for its use. The former led him to view causal
inference as a conditioned reflex. Only if ideas were associated in a
large number of cases, would the conditioning process be effective.
Constant conjunction was therefore a causal prerequisite of our
making causal judgements. It is also, in general, an adequate justification, and so was in his view the main constituent of the concept.
But although it is, indeed, an important aspect, it is neither a necessary nor a sufficient condition. We often in practice make causal

judgements on the basis of fairly few observations. Sometimes a single instance—as when we are determining the melting-point of a pure chemical—is enough; and although numbers of instances can be, as we shall see,[1] of value, for the most part we are concerned not with multiplying instances, but with delimiting conditions. We are guided rationally, by the canons of inductive argument, and do not need great numbers of instances to bludgeon our brains into expecting the effect whenever the cause has been observed. Moreover, counter-instances, when they are observed, do not have quite the fatal effect they should have, were constant conjunction our pre-eminent rule. Counter-instances are important, and may be taken, at a first approximation, as falsifying putative hypotheses. But although they always should be taken seriously, we find[2] that in some cases it is the counter-instance which is discounted—as being due, for example, to experimental error—rather than the causal generalisation; and therefore we should not make out that constant conjunction is a strictly necessary condition of there being a causal relation.

Nor is it a sufficient condition. A-type events may have been constantly conjoined with Z-type events, but it does not follow necessarily that A-type events therefore cause Z-type events. Day invariably follows night, but is not caused by it. If we come across a constant conjunction it is, indeed, likely that the first-occurring event is the cause of the second, but it may be that they are both the effects of some common cause, or are both manifestations of the same underlying causal process (as day and night are both manifestations of the diurnal rotation of the earth). And it is always possible, and a proper cause of worry to the philosopher, that an observed constant conjunction is not due to any causal relation at all, but is just a coincidence.[3]

Constant conjunction, thus, cannot be constitutive of our concept of cause. Although it is an important aspect it is not a fundamental aspect but only a derivative one. It is important because it is a usable and useful test of when we are entitled to assert that A is the cause of Z. But once we distinguish that question from the question of what it means to say that A is the cause of Z, and distinguish the analysis of the concept from the criteria for its application, we see

[1] See below, Ch. IV, p. 62.
[2] See above, Ch. I, pp. 4–5.
[3] See below, Ch. IV, pp. 66–7.

that the entirely extensionalist account offered by Hume misses part of its meaning. To say that A is the cause of Z is commonly taken to mean, as Hume himself admits, that A not only always comes before, or at least not after, Z, but in some way necessitates A, and this is felt to be something more than bare constant conjunction. We are not merely saying that A is *invariably* followed by Z, but that it is *inevitably* followed by Z. A causal relation, we believe, besides being weakly antecedent, is necessary as well as being in consequence repeatable. There is also, as I shall argue later, some idea of its being explicable, which manifests itself in particular in the requirement of what I shall term "contiuity". These five characteristic features of a causal relation, namely that it is

(1) necessary,
(2) weakly antecedent,
(3) repeatable,
(4) explicable,
(5) "contiuous",

are, as I shall attempt to show, interconnected, and are woven together to constitute a coherent concept.

Hume rejects the common view of what we mean when we say that A is the cause of Z, because he has difficulties with the concept of causal necessity. It is not that he finds the concept of necessity altogether unintelligible, for he does recognise what we should describe as logical necessity. To borrow the symbolism of modern modal logic, he not only sees the parallel between $\square$ and (x)—and, correspondingly, between $\diamond$ and ($\exists$x)—but also admits the rule

$$\vdash \Gamma \Rightarrow \vdash \square \Gamma$$

if Γ is a theorem of deductive logic, then it is not only true, but necessarily true.

But this is only one sort of necessity. Although we do not doubt the necessity of logically necessary propositions, there is little reason to suppose that only logically necessary propositions are in any sense necessary. We often talk of physical necessities, biological necessities, legal necessities, social necessities, moral necessities, etc., and all these locutions are intelligible and useful. We need to distinguish the internal logic of modal words from the particular modalities in which they operate. The words 'necessary', 'impossible' and 'possible' are mutually related by what we might call the triangle of opposition, by analogy with the square of opposition used to exemplify the relations between the quantifiers. Thus if p is necessary, then not-p is impossible, and if not-p is impossible, then it is not the case that not-p is possible, in just the same way as if all x are F, then no x are not F, and it is not the case that some x are not F. If we add non-necessity to our list of modal concepts, we can obtain an exact

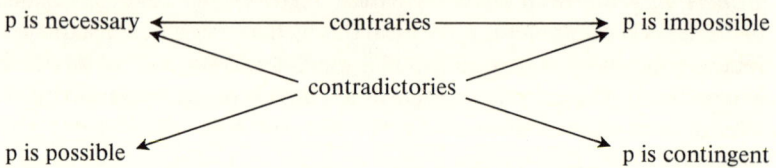

Note: The word 'contingent' is also used to mean 'both non-necessary, and possible'.

FIGURE 1

parallel of the traditional square of opposition, as in Figure 1. These relations hold whatever sort of modality we are operating with, and are used as much by the lawyer as the logician.

A sceptic could grant this much, and yet be unconvinced that there were any intelligible modalities other than the logical one. After all, he can introduce logical necessity by means of his rule

$$\vdash \Gamma \Rightarrow \vdash \Box \Gamma$$

whereas no similar □-introduction rule has been offered for any other sort of necessity. Hume felt himself driven to explain it away as being merely a sort of conditioned reflex because he did not see where we could have got it from. Constant conjunction expressed for him all that could be properly expressed of the notion of necessity we attribute to causality. The connexion between necessity and universality is profound, and Hume was quite right to take constant conjunction as a characteristic mark of necessity. But it is not clear that it is equivalent to it. Leibniz explained necessary propositions as being those that held in all possible worlds, but the word 'possible' is as modal as the word 'necessary'. Modern philosophers distinguish closed universal statements, such as 'All my friends have names beginning with P', from open ones which not only are, as it happens, true, but imply further counterfactual propositions, in the way that 'All men are mortal' implies 'If Zeus were a man, he would be mortal'. This, too, is an important point, but it again involves changes of mood which sit ill in Hume's flat world-view. If we confine ourselves to indicative statements of what actually is the case, we cannot actually render the modal force of necessity, and our account of causality will be correspondingly defective.

Although the necessity of a causal relation is not a logical necessity like that of a deductive entailment, and although it cannot be discerned by the senses, we are nonetheless able to form the concept. In fact we form it from two sources, partly from our practical experience of trial and error, partly from our theoretical understanding of the necessities of reasoning. We use causes. We are agents who decide to do things, and use causes to bring about the ends we desire. We cause things to happen, often by causing their causes to happen, and thus begin to form an idea of a cause operating in the natural world as a back-formation from our own causal interventions. Hume argues to the contrary,[4] but takes altogether a too passive view of man. Instead of starting, as Descartes did, with an isolated cogitator who in the course of his speculations can doubt the existence of everything except himself, the philosopher should say, rather, *ego ergo ago*, and see himself primarily as an agent who does things and learns by doing them. Hume did not pay sufficient attention to our practical interest in causality, or to the epistemological consequences of our role as agents. Although he called himself an Empiricist, he had a surprisingly un-empirical approach to knowledge. He did not recognise the importance of experiment, or of the fact that much of our knowledge is obtained by trial and error. Hume's account of human nature is that of a very passive percipient, not that of an active agent who learns a lot in the course of doing things. But it is chiefly by doing things that we discover causal connexions in the world around us, and in particular obtain a sense of what is possible and what is not. Our notion of empirical possibility is derived not from the compatibility or incompatibility of ideas, but from trying and failing to do things.[5] Trial and failure give us an empirical sense of impossibility, and once we have a notion of empirical impossibility, we have a notion of empirical necessity too. If, however hard we try, we can never have A without its being followed by Z, then we have a notion of Z's necessarily following upon A. Agents who intervene in the world around them, and attempt, not always successfully, to make nature conform to their own wishes, readily form the concept of natural necessity, and also have no difficulty in

---

[4] *Enquiry*, Sect. VII, part I, 52–3.
[5] Ch. II, p. 21.

distinguishing causal connexions from chance concomitances. The difference between a constant conjunction which is due to a causal connexion and one which is merely a coincidence can be discovered by our making an experiment and intervening to see if we can bring about the putative cause without the effects thereupon following. Any coincidence, or any pre-established harmony, would thereby be disrupted, and only those concomitances that were due to genuine causal connexions would be left. Experiments serve as a sort of probe which enables us to feel out the real shape of reality around us, as opposed to the flat appearance of superficial actuality. We are not confined to observing what, as it happens, actually occurs, but can find out also the sort of thing which would occur, were circumstances different, just because we can introduce adventitious differences of circumstance by means of our own arbitrary interferences. Instead of being merely observers who are confined to reporting in the indicative mood what actually has happened, we are agents who can use also the subjunctive mood to consider what would be the case if we were to intervene in such and such a manner. The distinction between closed and open universal propositions thus becomes real, and if we are talking in terms of natural law we are able to distinguish natural laws from accidental generalisations. And equally, if we are talking in terms of causes, we can give sense to their being four-dimensional analogues of switches, by means of which we can manipulate and control the course of events, and can therefore construe causal necessity as being a four-dimensional analogue of solidity.

Our practical concern with causes as means of manipulating things is of great importance. It guides us in our selection of that part of the antecedent sufficient condition which is most worth picking out as *the* cause. It also explains the second and third characteristics of the causal relation, weak antecedence and repeatability. If we want to know about causes in order to use them to bring about effects which we desire, then, since it is impossible to alter the past, it is not enough to know sufficient conditions of the effect we want to bring about. We shall need to know, rather, those conditions which are both sufficient and antecedent in time (in a weak sense which may approach the cause's being simultaneous with the effect but certainly excludes its being subsequent to it). There are many other sufficient conditions—such as the symptoms by means of which a doctor diagnoses what is wrong—but if they occur after the event in question, they are of no practical use to us in bringing it about.

Our practical concern with causes gives rise to the requirement not only of weak antecedence but of repeatability. If we are to be able to use causal relations in order to bring about Z, we must be able to know on each occasion what to do to cause Z. It would be no use if on one occasion the kettle's boiling was caused by putting it on the fire but on all others it was caused by quite different conditions—say a spell one time, dancing a ritual dance another, turning its spout towards the magnetic north a third time, and so on. Although there may be more than one way of bringing about Z, each way must be a *way*. We can boil a kettle by putting it on the fire, putting it on a gas ring, turning on the electricity, focussing the sun's rays through a large magnifying glass: but each of these methods works not just once only, but time and again. Else it would not be a method, and would be useless.

The practical approach to causality although important is not the only one. The word 'cause', as Collingwood pointed out,[6] had its original habitat in our discourse about human affairs, not natural phenomena. The Latin word *causa* was used primarily to refer to a reason for action. In politics we still sometimes talk about The Cause, and when we are button-holed by a Woman with a Cause, she presses on us reasons why we should join in her crusade for the realisation of some social good. The application of the word to natural phenomena is a derivative use. It does not follow, as Collingwood made out, that it is an improper use, but it does mean that the concept is more closely connected with that of explanation than Hume and his followers allow. By 'cause' we often mean explanation, characteristically introduced by the word 'because'—originally 'by cause' (= 'by reason of')—and the principle of sufficient reason, that Every Event has a Cause, meant that every event had an explanation, not that to every event was correlated some antecedent event such that events of the two types were constantly conjoined. In modern English usage the sense that is peculiarly appropriate to natural phenomena is expressed by the adjective 'causal', and it is sometimes useful to distinguish "causal" causes from those that operate in human affairs through the minds of men. But this distinction does not imply that causal causes are, as Hume maintained, completely opaque to the human understanding. If Hume were right, to say that smoking causes cancer would mean no more than to say that smoking and cancer are constantly conjoined, and the claim of

6 R. G. Collingwood, *Metaphysics*, Oxford, 1940, Ch. XXIX.

the tobacco companies that the correlation between smoking and cancer is not a causal one and is to be explained in some other way would be not merely untrue but unintelligible. Yet we can understand their claim, however false we reckon it to be. And so we find Hume's extensionalist account inadequate. We think that something more could be found out to establish that smoking causes cancer. Although the statistics are fairly damning, what would clinch the case would be the discovery of some "causal mechanism", some explanation of how nicotine or tar in the lungs made cells become cancerous. Causes are explanations, and although at some stage of scientific enquiry we may be unable to explain why A is invariably followed by Z, and may have to accept that there is a causal connexion even if we are unable to understand it, we as scientists seek explanations and will not rest content until we have been able to explain it to our own satisfaction.[7]

The commitment to explicability is less than absolutely clear. We do not know in advance what the rationale of a satisfactory explanation will be. Nevertheless, we do find some explanations deeply convincing, and reckon that some natural phenomena must for scientific reasons occur in the way they do. Thus the rationality of scientific causes is a second source for their being necessary. The requirement of repeatability can also be grounded in that of rationality. Rational explanations are, if not universal, at least universalisable. Even in human affairs, if I explain my actions, I am thereby committed to accepting, in the absence of some further differentiating circumstance, a similar explanation for a similar action on your part. Scientific explanations are subject to a comparable, only more stringent, requirement. In the case of human explanations we both need to have the let-out clause "in the absence of some further differentiating circumstance" on account of the infinite potentiality of human beings, and can afford to on account of our ability to "get inside" another man's skin, and weigh up his reasons from an agent's standpoint. If we attempt to lay down some complete code of conduct—as many systems of law have sought to do—there will always turn up new, unexpected cases, which the rules do not adequately cover. Often we can reason out what ought to be done, but our reasons will be additional to the existing corpus of law. We therefore need a safety-valve, and have to hedge our rules with some qualification, saying that they apply only "other things

[7] See further below, Ch. XI, pp. 176–7.

being equal". This, of course, opens the door to all sorts of abuse. People will often make out that their case is a special case. But we can detect and rule out many forms of special pleading on the part of others, because we are, like them, rational agents, at least to some minimal degree of rationality, and therefore know what it is like to act for a reason, and can, within limits, assess the reasons offered. No such understanding is possible in natural science. We cannot get inside an electron's skin, and consider the *pros* and *cons* of jumping from one quantum level to another. The idea is ludicrous. All our understanding of natural phenomena is necessarily external. But together with this restriction on our powers of understanding there is a corresponding limit on our need. Natural phenomena, although sometimes complex, do not have the infinite potentiality that human beings have. Although we may need to modify our scientific hypotheses and putative laws of nature from time to time, they do not need to admit indefinite modification. Indeed, it is the mark of a bad scientific hypothesis, which fails to express a real law of nature, that it is continually needing modification. With natural phenomena, only a finite number of factors are relevant. Things can be qualitatively identical but numerically distinct, for not every differentiating feature—in particular, not a spatial or a temporal difference *per se*—counts as a quality. Leibniz's principle of the Identity of Indiscernibles[8] notwithstanding, we accept the chemists' view of matter as made up of myriads of atoms, the atoms of any one isotope of any element being all alike, though differing in their spatio-temporal locations. Different leaves of a tree may differ in some significant respects, as do different specimens of the same species; but this we account for by reason of the complexity of the genetic material and the variability of the environment. In the last chemical analysis, every atom of hydrogen, or of carbon, or of oxygen, is the same in all significant respects as every other atom of that isotope of that element. Although the complexity of biological phenomena is formidable, our fundamental view of natural phenomena is of their being articulated into kinds or, as we shall call them, types, each type having potentially many instances. There may be—there often is—a problem of identifying the relevant factors and specifying the types adequately; but this task once accomplished, we shall find

[8] Leibniz, *Fourth Letter to Dr. Clarke*, § 4 in H. G. Alexander, ed., *The Leibniz-Clarke Correspondence*, Manchester UP, 1956, p. 36. See further below, Ch. VIII, pp. 129-32.

causal relations manifested, as Hume maintained, in constant conjunctions of instances of the relevant types. Repeatability is required not only for the causal relation to be usable, but as a mark of its claim to be, at least in principle, explicable and therefore rational.

Temporal and logical priority are entirely distinct. Rationality does not require temporal priority or antecedence. We cannot, therefore, account for all the characteristics of causality in terms of rationality alone, although we can, perhaps, account for the uncertainty we have felt about the importance of temporal antecedence. There are many scientific questions—what is the cause of copper sulphate's being blue? What is the cause of the sky's being blue? What is the cause of the earth's magnetic field?—where issues of temporal priority do not arise, or it seems awkward to insist on strict antecedence. In some cases we may decide that a different sort of explanation is being looked for, which ought to be distinguished from a causal explanation properly so called; in others, however, we may feel it forced to maintain that there must be a categorical distinction, and therefore we concede that some causes may be contemporaneous with their effects. And in such cases we regard the cause as a cause not because it is a means of producing the effect in question, but rather because it is a way of explaining it.

Although we cannot explain causality in terms of explicability alone, there is a further characteristic, apart from explicability itself, which can be explained only on the ground that causes ought to be susceptible of explanation. Hume himself bore unwitting witness to this both in his "Rules by which to judge of causes and effects",[9] the first of which laid down

1. The cause and effect must be contiguous in space and time

and in his earlier and more explicit gloss[10]

I find in the first place, that whatever objects are consider'd as causes or effects, are *contiguous*; and that nothing can operate in a time or place, which is ever so little remov'd from those of its existence. Tho' distant objects may sometimes seem productive of each other, they are commonly found upon examination to be link'd by a chain of causes, which are contiguous among themselves, and to the distant objects; and when in any particular instance we cannot discover this connexion, we still presume it

[9] David Hume, *A Treatise of Human Nature*, Bk. I, Part III, Sect. XV, p. 173.
[10] Bk. I, Part III, Sect. II, p. 75.

to exist. We may therefore consider the relation of CONTIGUITY as essential to that of causation . . .

Later I shall argue that Hume's contiGuity should really be construed as a requirement of contiNuity—or, rather, continuous connexion—and is of very great importance in the philosophy of physics. But in any case, the principle of contiuity as we may call it, not to prejudge the issue, or locality as it has been termed in recent discussions of quantum mechanics, is a principle commended to us by reason, not a fact forced on us by experience, and is not so much a rule for the application of the concept as part of the concept itself. If we have a sufficiently constant conjunction of two phenomena, we are prepared to accept it as good evidence for saying that the first is the cause of the second, even though they may be widely separated in space, and considerably separated in time. We do not need to assure ourselves in advance that there is some spatio-temporal continuity, but rather assume that there must be, once we are convinced that the two sorts of phenomena are causally related. We may think that action at a distance is conceptually impossible, as being contrary to reason, but it is not logically impossible, and there have been times when contemporary science has seemed to show that it actually occurs.[11] When Locke first asked himself how bodies produce ideas in us, he replied "manifestly by impulse and nothing else. It being impossible to conceive that body should operate on what it does not touch (which is all one as to imagine it can operate where it is not), or, when it does touch, operate any other way than by motion", from which he went on to argue "If, then, bodies cannot operate at a distance . . . it is evident that some motion must be thence continued by our nerves or animal spirits, by some part of our bodies, to the brains or the seat of sensation, there to produce in our minds the particular ideas we have of them". But after reading Newton's *Principia*, Locke acknowledged that gravitation did operate on matter at a distance, and changed the fourth edition of the *Essay*, to say that impulse was merely the only way we could conceive bodies to operate in.[12]

Although we may be forced by the evidence to acknowledge a causal relation which appears not to satisfy the requirement of contiuity, we cannot explain how it works until we have discovered some

[11] See below, Ch. IV, p. 58, Ch. VIII, p. 136, and Ch. XI, pp. 176-7.
[12] *Essay Concerning Human Understanding*, II, 8, §§ 11–12. See also *Reply to Second Letter*, p. 468; quoted in A. S. Pringle-Pattison's edition, p. 68n.

spatio-temporally contiuous path of causal influence linking the cause with the effect. Our concept of causality is therefore seen to be thus linked with our concepts of space and time, not simply on account of the meanings of the words, nor as something given us in experience, but as a requirement of reason. We have a view of space and time which rules out action at a distance as being essentially inexplicable, and therefore leads us to posit, even when we cannot actually discover, some contiuous connexion linking cause and effect.

Causality emerges from Hume's criticism not entirely unscathed, but still as a coherent and intelligible idea, although not always an absolutely clear and distinct one. The focus is slightly blurred because we develop the concept in both our practical and our theoretical reasoning. From our attempts to manipulate things we discover natural impossibilities, and hence natural necessities, and from this point of view see causes as antecedent in time and as repeatable. In our theoretical reasoning our prime concern is to explain things, and in so far as we succeed, or trust that we shall succeed, we see causes as rationally necessary, repeatable and contiuous. We can tie together the five characteristic features of a causal

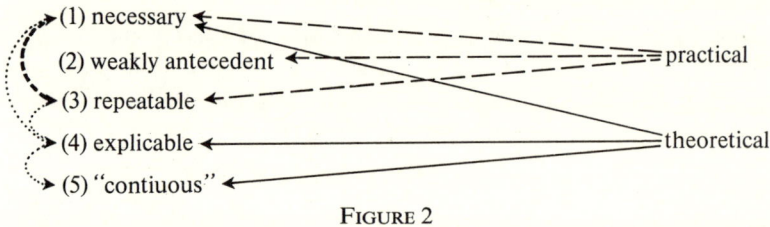

FIGURE 2

relation (1), (2) and (3) arise from our using the concept in our practical reasoning; (1) (4) and (5) from our using it in our theoretical reasoning. Independently of our underlying concerns, (1) and (4) each by themselves would yield (3); and (4) by itself yields (1), and, according to our usual way of thing about natural phenomena, (5).

Further Reading

B   Bertrand Russell, "On the Notion of Cause", *Proceedings of the Aristotelian Society*, xiii, 1912-13, pp. 1-26; reprinted in *Mysticism and Logic*, London, 1918, pp. 180-208, Pelican, 1953, pp. 171-96.

B   R. G. Collingwood, *Metaphysics*, Oxford, 1940, Part IIIc, Chs. XXIX-XXXII, pp. 285-327.

B   David Hume, *A Treatise of Human Nature*, Book I, Part III, Sections XIV and XV, pp. 155-76).

For a careful, but difficult, critique of Hume's theory of causality, see J. L. Mackie, *The Cement of the Universe*, Oxford, 1980, Ch. I.

B   J. S. Mill, *System of Logic*, Book III, Chs. 8–10.

IV

The Logic of Cause and Effect

THE two questions 'What does it mean to say that A is the cause of Z?' and 'Under what conditions is one entitled to say that A is the cause of Z?' are, we have maintained, distinct. But they are related. And, now that we have completed our analysis of the concept of cause, we need to consider the criteria for its application.

Hume gives eight rules by which to judge of causes and effects:[1]

1. The cause and effect must be contiguous in space and time;
2. The cause must be prior to the effect;
3. There must be a constant union betwixt the cause and the effect;
4. The same cause always produces the same effect, and the same effect never arises but from the same cause;

together with four further, derivative rules, which we may regard as a crude anticipation of Mill's methods of agreement and difference. Rules 1 and 2 we have already discussed. We defer discussions of the second half of rule 4 until later.[2] Rule 3 and the first half of rule 4 both express the requirement of constant union, or constant conjunction as Hume often puts it, which is equivalent to our requirement of repeatability. It is of central importance when the question 'Under what conditions is one entitled to say that A is the cause of Z?' is raised, because it is eminently a condition we can check up on. We can test in a particular instance whether or not A is conjoined with Z. If it is not, we have falsified the putative law

$$(x)[A(x) \rightarrow Z(x)]$$

---

[1] David Hume, *A Treatise on Human Nature*, Part III, Sect. 15, pp. 173–4. These rules were subsequently developed and refined by J. S. Mill in his *System of Logic*, Book III, Chs. 8–10. For a modern exposition, G. H. von Wright, *A Treatise on Induction and Probability*, London, 1961, Ch. IV. J. L. Mackie, *The Cement of the Universe*, Oxford, 1974, Appendix, pp. 297–321.

[2] P. 67 and Ch. X, p. 158.

in accordance with Popper's precepts:[3] if A is in the particular instance conjoined with Z, then we have not proved that A is always conjoined with Z, but have left open the possibility that it may be. In order to be sure that a particular causal factor A is the cause of Z, granted that there is some causal law of the form

$$(x)[F(x) \rightarrow Z(x)],$$

we need to *eliminate* other possibilities equally plausible at first sight. If there are causal laws, and if we can consider all the possible causal laws, and if we can eliminate all the other possible causal laws, then we can be reasonably sure that A *is* the cause of Z. But they are big IFs. The biggest we have already discussed in Chapter II. In this chapter we identify the other difficulties we need to overcome and assumptions we need to make, in order to be able to reach the conclusion that a putative causal law does indeed hold.

The first difficulty is that instances are not pure instances of just one type but always instances of many different types. I get ill the day after having eaten oysters when there was no R in the month; but it is also the day after having rowed in the second eight, having drunk more port than I am used to, having had a chilling reception of my essay, having been given the brush off by Penelope, having visited Tom who had got gastric flu, etc., etc. I cannot be sure, on the basis of this one incident, that it was the oysters which were responsible. It may have been the rowing, it may have been the port, it may have been my tutorial, Penelope's personality or Tom's germs. Obviously, however, each of these hypotheses is open to falsification. I have rowed before without being ill, so it cannot be just the rowing; similarly, I have drunk port, had unsatisfactory tutorials, had my amorous ambitions abruptly disappointed, been exposed to infection, without suffering a digestive upset, so I can eliminate those simple hypotheses. Given any particular causal hypothesis, I can look round for evidence against it, or devise an experiment to put it to the test. If, for example, the original instance was one where possible causal factors A_1 and A_2 were present, and you counter my suggestions that the result, Z, was the effect of A_1, by suggesting instead that it was the effect of A_2, we look round for, or try to bring about, instances of A_1 without A_2 and of A_2 without A_1. It may be that Z results in the one case and not in the other. If Z occurs when we have A_1, then we can rule out your counter-suggestion that it

[3] Ch. I, p. 6.

was A_2 which caused Z. If you suggest that my illness was due to the cigar I smoked rather than the oysters, and if I value knowledge of causes more than a quiet stomach, I can try next week smoking a cigar without eating oysters, and the following week eating oysters without smoking a cigar. If I am not ill next week, then it cannot be just smoking cigars that makes me ill, and if I have another queasy turn the following week, then oysters are still under suspicion.

The argument is crude, but effective. We have a number of rival hypotheses,

$(x)[A_1(x) \rightarrow Z(x)]$
$(x)[A_2(x) \rightarrow Z(x)]$
$(x)[A_3(x) \rightarrow Z(x)]$ etc.,

and a number of individual instances,

$A_1(1)$ & $A_2(1)$ & $A_3(1)$ & $Z(1)$
$A_1(2)$ & $A_2(2)$ & $A_3(2)$ & $Z(2)$
$\sim A_1(3)$ & $A_2(3)$ & $\sim A_3(3)$ & $\sim Z(3)$
$\sim A_1(4)$ & $\sim A_2(4)$ & $A_3(4)$ & $\sim Z(4)$ etc.

We eliminate the second and third rivals on the strength of the third and fourth instances. Formally we argue that if, say, the second hypothesis had been true, then, by the rule of inference called instantiation or universal elimination,

$A_2(3) \rightarrow Z(3)$

which, by propositional calculus, is equivalent to

$\sim(A_2(3)$ & $\sim Z(3))$.

But from the third instance, there follows, by &-elimination,

$A_2(3)$ & $\sim Z(3)$.

So the second hypothesis is inconsistent with the third instance, and if we have observed the third instance, the second hypothesis is ruled out. Similarly, the third hypothesis is ruled out by the fourth instance. Thus, as between those three rival hypotheses, the observed instances rule out the second and third, and to that extent support the first.

EXAMPLE I

Simple Elimination of Simple Hypotheses

Assumption I. All causal laws are of the simple form $(x)[A(x) \rightarrow Z(x)]$

Observed Phenomenon	$A_1(1)$ & $A_2(1)$ & $A_3(1)$ & $Z(1)$. I ate oysters, smoked a cigar, had too much port, and was ill.
Suggested Explanation	$(x)[A_1(x) \rightarrow Z(x)]$ Oysters make me ill. This fits the fact that last term too I ate oysters, smoked a cigar, had too much port, and was ill. $A_1(2)$ & $A_2(2)$ & $A_3(2)$ & $Z(2)$.
Rival Explanations	(i) $(x)[A_2(x) \rightarrow Z(x)]$ Cigars make me ill. (ii) $(x)[A_3(x) \rightarrow Z(x)]$ Port makes me ill.

Assumption II. It must have been either the oysters or the cigar or the port that made me ill.

Rival Explanation (i) $(x)[A_2(x) \rightarrow Z(x)]$ Cigars make me ill.	inconsistent with	Instance $A_2(3)$ & $\sim Z(3)$ I smoke a cigar next week without being ill.
Rival Explanation (ii) $(x)[A_3(x) \rightarrow Z(x)]$ Port makes me ill.	inconsistent with	Instance $A_3(4)$ & $\sim Z(4)$ I drank port last week without being ill.

So

Oyster hypothesis, alone of those put forward, remains in the field.

The argument is often criticized as being negative rather than positive: we refute the alternatives, but do not conclusively prove the truth of the remaining hypothesis. It always might be the case that the real cause is something quite different. That is true, but is a ground for caution rather than scepticism. Unmade suggestions are difficult to refute, but difficult to take seriously either. If the sceptic confines himself to pointing out that other hypotheses, besides the favoured one, are compatible with the evidence, we listen politely, and pass on. We knew that. Once we realised that inductive arguments were not deductive, we knew that the evidence was consistent with the negation of the conclusion, and would be likely to be consistent with many propositions having the form of a universal implication. In order to be taken seriously the sceptic must not merely tell us that inductive arguments are not deductive, but put forward an alternative hypothesis. Once he does that, we can put it to the test. He may be proved right, he may be proved wrong. Either way, we can advance our knowledge by ruling out some hypothesis, and narrowing down the range of possible causal factors. But if the sceptic does not offer any alternative explanation, we do not pay much attention. We are seeking explanations, and believe that there is some causal explanation of the phenomenon. Our proposed explanation may not be the right one, but, until a better one is offered, it is the best we have.

The counter to the sceptic is adequate as against the persistent sceptic, but does not answer all our doubts. The simple argument by elimination will decide between rival hypotheses of the form

$$(x)[A_1(x) \rightarrow Z(x)],$$

but many causal hypotheses are not as simple as that. For one thing, it is often the combination of two or more causal factors that is effective—the match lights only if it is struck *and* there is oxygen present. For another, negative factors are as important as positive ones—the match lights only if it is *not* wet. Thirdly there can be more than one cause of the same event. Arsenic causes death. So does cyanide. We have the possibility of there being two or more causal laws, of the form

$$(x)[A_1(x) \rightarrow Z(x)]$$
$$(x)[A_2(x) \rightarrow Z(x)].$$

These can be combined into a single law of the form

$$(x)[A_1(x) \lor A_2(x) \rightarrow Z(x)].^4$$

---

[4] This is a simple logical equivalence, although not an immediately obvious one. It is a theorem of the propositional calculus that

$$[(p \rightarrow r) \& (q \rightarrow r)] \leftrightarrow [p \lor q \rightarrow r]$$

Where we have a conjunction of causal hypotheses we are hypo-
thesizing a disjunction of causes. Either arsenic or cyanide causes
death. We therefore need to consider not just simple hypotheses of
the form

$$(x) [A_1(x) \rightarrow Z(x)]$$
$$(x) [A_2(x) \rightarrow Z(x)] \text{ etc.,}$$

but complex hypotheses, where the antecedent is a disjunction of
conjunctions of causal factors being either present or absent. If we
write negations as $\overline{A_1}(x)$ instead of $\sim A_1(x)$, etc., a causal hypothesis
will be of the general form [5]

$$(x)[A_1(x) \& A_2(x) \& \overline{A_3}(x) \lor A_1(x) \& \overline{A_2}(x) \& A_3(x) \rightarrow Z(x)] \quad \text{I}$$

If I know what the possible causal factors are, I can set about
testing each conjunction of causal factors being either present or
absent. Suppose the possibly relevant causal factors are A_1, A_2,
..., A_r; there are 2^r combinations of these factors being either present or absent, which we could tabulate like a truth-table:

$$A_1 \& A_2 \& \ldots \& A_r$$
$$\overline{A_1} \& A_2 \& \ldots \& A_r$$
$$A_1 \& \overline{A_2} \& \ldots \& A_r \quad \text{II}$$
$$.\quad.\quad.$$
$$.\quad.\quad.$$
$$\overline{A_1} \& \overline{A_2} \& \ldots \& \overline{A_r}$$

Each of these combinations excludes every other one. Each instance
could be classified as exemplifying one, and only one, such combination. We could, although it might be a lengthy and tedious
process, go through them one by one, eliminating those in which Z
did not ensue, and retaining any in which it did. We shall end up
with a causal hypothesis where the antecedent is in disjunctive normal form, containing one or more disjuncts, each of which is a
conjunction of r terms, A_1, A_2, ... A_r, or their negations.

Disjunctive normal form is cumbersome. It is often possible to
simplify it. In the example that follows we shall suppose that in fact
the cause of Z is the combination of A_2 and A_3, although at first A_1

[5] For the sake of simplicity, brackets are omitted according to the standard convention, whereby the connectives $\sim$, $\&$, $\lor$, $\rightarrow$ are taken, in the absence of brackets,
as binding in that order. If brackets are put into I, it reads

$$(x) [((A_1(x) \& A_2(x) \& (\sim A_3(x))) \lor (A_1(x) \& (\sim A_2(x)) \& A_3(x))) \rightarrow Z(x)].$$

was the prime suspect. A detailed examination of each of the 2^r possible combinations will eliminate all except the following 2^{r-2} cases, in which every characteristic may be present or absent, except for A_2 and A_3, which are always present:

A_1 & A_2 & A_3 & A_4 & ... & A_r
$\overline{A_1}$ & A_2 & A_3 & A_4 & ... & A_r
A_1 & A_2 & A_3 & $\overline{A_4}$ & ... & A_r
$\overline{A_1}$ & A_2 & A_3 & $\overline{A_4}$ & ... & A_r

· · ·

A_1 & A_2 & A_3 & A_4 & ... & $\overline{A_r}$ III
$\overline{A_1}$ & A_2 & A_3 & A_4 & ... & $\overline{A_r}$

· · ·

· · ·

A_1 & A_2 & A_3 & $\overline{A_4}$ & ... & $\overline{A_r}$
$\overline{A_1}$ & A_2 & A_3 & $\overline{A_4}$ & ... & $\overline{A_r}$

The total sufficient condition is represented by the disjunction of all these 2^{r-2} cases, any one of which is a sufficient condition. We could write it thus:

A_1 & A_2 & A_3 & A_4 & ... & A_r v $\overline{A_1}$ & A_2 & A_3 & A_4 & ... & A_r v
A_1 & A_2 & A_3 & $\overline{A_4}$ & ... & A_r v $\overline{A_1}$ & A_2 & A_3 & $\overline{A_4}$ & ... & A_r v

· · ·

A_1 & A_2 & A_3 & A_4 & ... & A_r v A_1 & A_2 & A_3 & A_4 & ... & $\overline{A_r}$ IV

This could be simplified by the standard methods for obtaining conjunctive normal forms. We can write the first two conjunctions as a product

$(A_1$ v $\overline{A_1})$ & $(A_2$ & A_3 & A_4 & ... & $A_r)$

and similarly the second two

$(A_1$ v $\overline{A_1})$ & $(A_2$ & A_3 & $\overline{A_4}$ & ... & $A_r)$

and, putting them all together, we have the whole in the form

$(A_1$ v $\overline{A_1})$ &
$(A_2$ & A_3 & A_4 & ... & A_r v A_2 & A_3 & $\overline{A_4}$ & ... & A_r v ...
... v A_2 & A_3 & $\overline{A_4}$ & ... & $\overline{A_r})$ V

Applying the same argument to A_4 we have

$(A_1 \vee \overline{A_1})$ & $(A_4 \vee \overline{A_4})$ &
$(A_2$ & A_3 & A_5 & ... & $A_r \vee A_2$ & A_3 & $\overline{A_5}$ & ... & $A_r \vee$ ...
... $\vee A_2$ & A_3 & $\overline{A_5}$ & ... & $\overline{A_r})$

Repeating for $A_5, A_6 \ldots A_r$, we obtain finally

$(A_1 \vee \overline{A_1})$ & $(A_4 \vee \overline{A_4})$
& $(A_5 \vee \overline{A_5})$ & $(A_6 \vee \overline{A_6})$ & ... & $(A_r \vee \overline{A_r})$ & A_2 & A_3 VI

Each of the disjunctions $A_1 \vee \overline{A_1}$, $A_4 \vee \overline{A_4}$, etc. are vacuous characteristics. An instance must possess the characteristic $A_1 \vee \overline{A_1}$ which
holds tautologically in every case. It adds nothing to say of an event
of any type that it is also of the $(A_1 \vee A_1)$-type, and hence one might
as well not say it. To say VI is no more than to say

A_2 & A_3 VII

and we may accept VII as tantamount to III, only in a more concise
form.

We can construe this as a new method of elimination: logicians
might like to call it the method of addition. Put crudely it runs: any
factor the presence or absence of which makes no difference to the
effect is irrelevant, and may be eliminated from the cause of that
effect. More precisely we may say: for any set of characteristics
that is a sufficient condition of a certain sort of event, if a set of
characteristics that is the same in all respects save one, where there
is the contradictory characteristic to that in the first set, is also
a sufficient condition of the same sort of event, then neither the
characteristic nor its contradictory are components of some cause
of that sort of event. Thus when boiling a kettle we may have been
always in the habit hitherto of saying a spell over it at the same time
as putting it on the fire: we are now wondering whether it is the
saying of the spell or the putting on the fire or both of them together
that is causing the kettle to boil. We can eliminate the saying of
the spell as a sole cause by trying it without putting the kettle on the
fire, and discovering that it does not boil. We can eliminate the
saying of the spell as a part-cause by trying the experiment of not
saying the spell but in every other respect behaving as before—i.e.
putting the kettle on the fire—we discover that the kettle boils just
as well as before, and correctly conclude that the saying of the spell
was causally irrelevant to the boiling of the kettle. Which is sound
common sense.

EXAMPLE II

Simple Elimination of Complex Hypotheses

Assumption I. Causal laws may be of the form

$$(x) [A_i(x) \& A_j(x) \text{ v } A_k(x) \& A_l(x) \rightarrow Z(x)]$$

Observed $A_1(1) \& A_2(1) \& A_3(1) \& Z(1)$

phenomenon I ate oysters, smoked a cigar, had too much port, and was ill.

Suggested $(x) [A_1(x) \& A_2(x) \& A_3(x) \text{ v } \overline{A_1}(x) \& A_2(x) \& A_3(x) \rightarrow Z(x)]$

explanation   Cigars and port together, with or without oysters, make me ill.

This fits the fact that last term too I ate oysters, smoked a cigar, had too much port, and was ill.

$A_1(2) \& A_2(2) \& A_3(2) \& Z(2),$

Assumption II. It must have been either the oysters or the cigars or the port, or some combination of them, that made me ill.

There are therefore eight possible combinations, each of which can be put to the test

1

$(x) [A_1(x) \& A_2(x) \& A_3(x) \rightarrow Z(x)]$ $A_1(1) \& A_2(1) \& A_3(1) \& Z(1)$

consistent

Oysters, cigars and port   with $A_1(2) \& A_2(2) \& A_3(2) \& Z(2)$

make me ill.

I have tried it twice with disastrous results.

2

$(x) [\overline{A_1}(x) \& A_2(x) \& A_3(x) \rightarrow Z(x)]$ $\overline{A_1}(5) \& A_2(5) \& A_3(5) \& Z(5)$

consistent

Cigars and port without   with   After working out example I,

oysters make me ill.   I suspected the oysters, so the following week I had cigar and port without oysters, but again with disastrous results.

3

$(x) [A_1(x) \& \overline{A_2}(x) \& A_3(x) \rightarrow Z(x)]$ $A_1(6) \& \overline{A_2}(6) \& A_3(6) \& \overline{Z}(6)$

inconsistent

Oysters and port without   with   I eat some oysters and drink

cigars make me ill.   some port but don't smoke next term, and am not ill

4

$(x)[\overline{A_1}(x) \& \overline{A_2}(x) \& A_3(x) \rightarrow Z(x)]$ $\overline{A_1}(4) \& \overline{A_2}(4) \& A_3(4) \& \overline{Z}(4)$

inconsistent

Port alone makes me ill.   with   I drank some port last week

(as in Example I)

5

$(x)[A_1(x) \& A_2(x) \& \overline{A_3}(x) \to Z(x)]$

Oysters and cigars make me ill.

inconsistent with

$A_1(7) \& A_2(7) \& \overline{A_3}(7) \& \overline{Z}(7)$

without having either oysters or cigars, and was not ill.

I eat some oysters and smoke a cigar but don't drink any port next term, and am not ill.

6

$(x)\,[\overline{A_1}(x) \& A_2(x) \& \overline{A_3}(x) \to Z(x)]$

Cigars without oysters or port make me ill.

inconsistent with

$\overline{A_1}(3) \& A_2(3) \& \overline{A_3}(3) \& \overline{Z}(3)$

I smoked a cigar but didn't eat any oysters or drink any port last vac, and wasn't ill.

7

$(x)\,[A_1(x) \& \overline{A_2}(x) \& \overline{A_3}(x) \to Z(x)]$

Oysters alone make me ill.

inconsistent with

$A_1(8) \& \overline{A_2}(8) \& \overline{A_3}(8) \& \overline{Z}(8)$

After my experience in 2 above I decide it is safe to have oysters by themselves, and eat some without ill effects.

8

$(x)\,[\overline{A_1}(x) \& \overline{A_2}(x) \& \overline{A_3}(x) \to Z(x)]$

Deprivation makes me ill.

inconsistent with

$\overline{A_1}(9) \& \overline{A_2}(9) \& \overline{A_3}(9) \& \overline{Z}(9)$

I have often been deprived of oysters, cigars and port without being ill.

So the only hypotheses consistent with the evidence are **1** and **2**, i.e.

$(x)[[A_1(x) \& A_2(x) \& A_3(x) \to Z(x)]$
$\& \,[\overline{A_1}(x) \& A_2(x) \& A_3(x) \to Z(x)]]$.

Oysters and cigars and port make me ill and cigars and port without oysters make me ill

which is the same as

$(x)\,[[A_1(x) \& A_2(x) \& A_3(x) \ v \ \overline{A_1}(x) \& A_2(x) \& A_3(x)] \to Z(x)]$

Either oysters and cigars and port or cigars and port without oysters make me ill

which is the same as

$(x)\,[A_2(x) \& A_3(x) \& [A_1(x) \ v \ \overline{A_1}(x)] \to Z(x)]$

Cigars and port, with or without oysters, make me ill.

which is the same as

$(x)[A_2(x) \& A_3(x) \to Z(x)]$

Cigars and port make me ill.

We thus have two methods of elimination, elimination by counter-example, and elimination by addition. The former refutes the suggestion that a particular combination of factors is a cause by showing that it is not in fact followed by the effect: the latter refutes the suggestion that a particular single feature is part of a causal complex by showing that it makes no difference to the causal efficacy of the rest of the complex. In our supposed example, we can imagine that the first instance was one in which A_1, A_2, A_3 and A_4 were all present, and A_1 was initially thought of as the cause. That hypothesis was refuted by counter-example—perhaps in the fifth instance A_1 was present, but not A_2 nor, as it turned out, Z. Subsequent instances in which first A_1 and then A_4 were absent, would show them both to be causally irrelevant to A_2 & A_3's being a cause. Thus we can pare down a number of possible causal factors to one or more combinations each of which is *a minimum sufficient condition* for the occurrence of Z.

If a combination of factors constitutes a minimum sufficient condition, anything less than that combination will not be a sufficient condition at all. Each constitutent factor of a minimum sufficient condition is therefore essential to its being a sufficient condition at all, or as we could say, necessary. Lawyers often distinguish necessary conditions, *conditiones sine qua non*, from the chief cause, *causa causans*, but it is best to avoid the use of the word 'necessary', as it invites confusion: besides its use here for an essential part of a minimum sufficient condition, it is used both as the correlative of a sufficient condition—if the cause is a sufficient condition then the effect is a necessary condition—and to express the necessity of the causal connection. An essential ingredient of a minimum sufficient condition is termed by Mackie "an insufficient but non-redundant part of an unnecessary but sufficient condition" or an "*inus*" condition for short.[6] Each essential ingredient can be regarded as the cause, if the other essential ingredients already obtain. Which essential ingredient is most naturally regarded as *the* cause, and which are regarded as only *conditiones sine qua non*, depend on a number of considerations—which is the most controllable,[7] which is the most unusual, sometimes, in law, which is the most culpable.[8] In science, however, we do not need to distinguish one essential ingredient as

[6] J. L. Mackie, *The Cement of the Universe*, Oxford, 1980, p. 62.
[7] R. G. Collingwood, *Metaphysics*, Oxford, 1940, p. 286 and Ch. XXXI.
[8] H. L. A. Hart and A. M. Honoré, *Causation in the Law*, Oxford, 1959.

being pre-eminently the cause. We are content to identify the whole complex of factors which together constitute a minimum sufficient condition, and the aim of our eliminative procedures is to find which among a set of possible causal factors are members of a minimum sufficient condition.

In practice we cut corners. We do not consider each combination separately, and then use the method of addition to simplify the result, but consider whether one factor, say A_3, is relevant to another's, say A_2's, being part cause. A_2 is thought of as a likely cause, but we are not sure whether it is effective on its own, or only in combination with A_3. We try the combinations $A_2 \& \overline{A_3}$, and find that no Z ensues; so it cannot be A_2 on its own. It might, however, be A_3 on its own, and so we test that hypothesis by seeing what happens if we have $\overline{A_2} \& A_3$. If in this case, too, no Z ensues, we know that A_2 and A_3 are causally relevant to each other. We then test $A_2 \& A_3 \& A_4$ and $A_2 \& A_3 \& \overline{A_4}$ to see if A_4 is relevant to the combinations of A_2 and A_3. In our supposed example, we find Z occurring in both cases, and conclude that A_4 is irrelevant. We can similarly test A_5, A_6, and any other factors we suspect of being causally relevant. We may find that none of them is relevant to A_2 and A_3's being causes, but it could still be the case that some factors in some combinations, say A_7 and A_8, are both causally irrelevant *to* A_2 and A_3's being causes, but may be causally relevant on their own. We need to distinguish the relation *being causally relevant to* from the predicate *being causally relevant*. In practice, however, we are often content to pick out only the more important causes, identifying exactly what other factors are causally relevant to them, without checking separately all the other logically possible combinations. We can justify this, logically speaking, by viewing a causal hypothesis with a disjunctive antecedent as a conjunction of separate causal hypotheses. If the whole truth is

$$(x)[A_2(x) \& A_3(x) \lor A_7(x) \& A_8(x) \to Z(x)]$$

we can say, with equal truth

$$(x)[A_2(x) \& A_3(x) \to Z(x)] \& (x)[A_7(x) \& A_8(x) \to Z(x)],$$

and can say that the combination of cigars and port is a cause of

subsequent illness, without prejudice to there being other combinations of factors—visiting ill friends without being resistant to
infection, say—which also cause illness. Often we are not wanting a
complete account of all the alternative causes of an event, but only
to pick out from the possible causal factors present on a particular
occasion those which were the actual causes. We alter the possible
causal factors one by one, and see what happens. If it makes a
difference, and there is no longer the same effect, then we can eliminate the other factors as constituting by themselves a minimum
sufficient condition and so conclude that the altered factor is an
essential ingredient of a minimum sufficient condition: if it makes
no difference, then we eliminate the altered factor as being causally
irrelevant to the combination of the unaltered factors. Thus if we
have r possible causal factors, we may need to perform only r tests,
not 2^r, in order to pick out from among them a minimum sufficient
condition. The saving in time is very great, and usually worth the
risk of not detecting an alternative cause. But when a car's not
starting is due both the damp in the ignition and to dirt in the
carburettor, short-cuts yield misleading diagnoses.

Thus far we have been assuming that we know what the possible
causal factors are. Often we have a good idea, based on our previous
experience and scientific theory, and against that background, the
procedures of elimination work well. But we could be wrong. It is
always possible that there is some factor that we had overlooked—
say the conjunction of Jupiter with Venus in the azimuth. We therefore relax the assumption that there are only finite number of known
possible causal factors, and consider how far, and with what justification, we can identify causes when we do not know that only certain
factors can be causally relevant.

We can still test for the causal relevance of any particular factor,
by altering it, and seeing what happens. If the effect does not then
ensue, the other causal factors present on the second occasion, whatever they are, do not together constitute a minimum sufficient condition. *If*, furthermore, we can assume that these other causal factors
are the same in the two cases, and that there was not some other
cause operating in the second case which was not operating in the
first, we can conclude that the altered factor is an essential ingredient
of a minimum sufficient condition which had been operative in the
first case but not in the second. So too, if the alteration of a factor
makes no difference, then, *provided* we can assume that the other

causal factors are the same in the two cases, we can conclude that the altered factor is causally irrelevant to whatever minimum sufficient condition it was that was causing the effect. We could in this way test as many factors as we wanted and with each discover either that it was an essential ingredient of a minimum sufficient condition or that it was causally irrelevant to the other factors. So, even if we do not know what all the possibly relevant factors are, we can still make progress, so long as we have a criterion of two cases being-the-same-in-all-other-relevant-respects. The situation is much like that in the mathematics of transfinite numbers; we cannot count a transfinite cardinal to tell how large it is, and whether it is of the same potency as some other, but we can tell that two cardinals are of the same potency without counting, by putting them into a one-one correspondence with each other.[9] More generally, the concept of 'being of the same potency as . . .' is logically prior to that of 'being of such and such a number', and similarly the concept of having all other relevant characteristics in common is prior to that of having in common all the specified relevant characteristics. Thus we can, granted some principles of causal irrelevance, prove a particular factor relevant or irrelevant to a certain causal connnexion, without having previously proved irrelevant all—possibly an infinite number—of the factors other than a specified and finite few.

We have, to begin with, the three presumptions of causal irrelevance we have noted already[10] that a difference of time, a difference of space and difference in state of mind of a human being are all, *per se*, irrelevant. We make the second more precise: not only a difference of position, but a difference of direction, or spatial orientation as we shall call it, must be *per se* irrelevant. A fourth presumption of irrelevance is that of the relative unimportance of the remote. We do not believe, except on very strong evidence, that the precise positions of the planets at the time of a man's birth will affect the course of his life. We may call this the anti-astrological canon. It follows from the principle of locality, or contiuity as we called it in the last chapter.[11] Any causal influence of distant events will be mediated through happenings in the locality, which, if they are going to affect anything, are likely to have other effects besides

---

[9] B. Russell, *Introduction to Mathematical Philosophy*, London, 1919, Ch. VIII; Leibniz, V §47, p. 71. See further below, Ch. VIII, p. 131.

[10] Ch. II, p. 22.

[11] Pp. 40–1.

the phenomenon under investigation, and are therefore likely to be detected. Although it is just possible that some distant circumstance, such as a conjunction of Venus with Jupiter, might have a highly specific effect through entirely straightforward means—perhaps through some resonance phenomenon—on an experiment on earth, if it did so, the means whereby it did so would be detectable and detected. It is not inconsistent to ascribe a causal influence to distant events without any intervening processes that might be detected, but we are rationally reluctant to do so. The strength of our reluctance can be gauged by the difficulties physicists have encountered when their theories have required action at a distance. We have already seen the difficulty Locke had in accepting Newton's theory of gravitation.[12] It was not only Locke. Newton himself was unhappy about action at a distance.[13] Until the success of Newton's system stilled all criticism, his own discomfort about gravitation was shared by many of his contemporaries. Leibniz thought that the only possible sort of mechanism was action by contact: "A body is never moved naturally, except by another body which touches it and pushes it. . . . Any other kind of operation on bodies is either miraculous or imaginary" (Fifth letter to Clarke, §34, Alexander, p. 66). Euler regarded it as "very strange and contrary to the dogmas of metaphysics" to ascribe causal influence to distant stars, and this remains a telling objection to Mach's thesis.[14] In quantum mechanics Einstein, Podolski and Rosen produced an argument to show the incompleteness of the quantum-mechanical description of reality. An essential strand of the argument was that any alternative explanation violated the principle of locality, that is, attributed consequences at one place to events at another without any intervening propagation of causal influence. Recent work on the foundations of quantum mechanics shows that physicists are prepared to go to considerable lengths rather than have to admit non-local causes.[15]

[12] Ch. III, p. 4.
[13] See below, Ch. XI, pp. 176–7.
[14] Quoted by D. W. Sciama, *The Unity of the Universe*, London, 1959, p. 96. See further below, Ch. VIII, p. 136.
[15] For a brief account of the Einstein–Podolski–Rosen Paradox, see H. Erlickson, "The EPR Paradox", *Philosophy of Science*, **39**, 1972, pp. 83–5; or, more fully, Max Jammer, *The Philosophy of Quantum Mechanics*, New York, 1974, Ch. 6. For recent developments see, e.g. H. P. Stapp, *Physical Review*, 1971, D3, pp. 1303ff., or M. L. G. Redhead, "Causality, Relativity and the EPR Paradox", *Foundations of Physics*, forthcoming.

These presumptions can all be justified as concomitants of the principle of finite causal relevance which we have already noted.[16] Natural science seeks general explanations. It does not admit of special cases, in the way in which the humane disciplines do. Causes have to be in principle repeatable, and thus universalisable in a strong sense. If a difference of position in space, or in time, or of some human being's state of mind, or of some distant part of the universe, were enough to make two situations relevantly different, then every situation would be different, and science would be impossible. It is a necessary presupposition of scientific enquiry, because it is a necessary condition of the applicability of the concept of cause and the possibility of scientific explanation, that two situations can be alike in all relevant respects. Hence we cannot cite a difference in spatial or temporal position or mental attitude or distant circumstance as by themselves making two situations dissimilar. Of course, this does not mean that they must be the same. If I measure the humidity in Cambridge, I may well get a different reading, because the climate is different there. The ether drift might well have had a different value in September from what it had in March, because the earth's orbital velocity was different. If my state of mind leads me to try an experiment with some of the conditions altered, it may well result in a different outcome. The sun is causally relevant to everything we do and experience on earth, and every time a man looks at the stars his nervous system is in a different state from what it would have been had the stars been differently disposed. But in each case we attribute the difference to some other factor than a bare difference of spatial or temporal position, spatial orientation, or mental attitude or distant circumstance. It may yet be objected, however, that where we do not know what the causal factors are, we cannot know that they are not varying between one situation and another which has a different spatial or temporal position, is differently regarded by some human being, or is subject to different influences from afar. In part our answer is the same as we gave earlier to the sceptic, that we shall consider these possibilities when they are substantiated, and an actual factor suggested as being causally relevant. But we can also say a little bit about the unknown. It is difficult for the unknown to vary very much with small changes of spatial position or very quickly with time, or to keep in step with the arbitrary and unpredictable interventions of the experimenter,

[16] Ch. III, pp. 39–40.

or to affect one distant region of space or time very much more than others near to that distant region. If I do two experiments in the same laboratory, or repeat one in the same position after only a short interval, the presumption that the two are the same in all relevant respects save those I have taken account of is well founded. It is still only a presumption, and could be rebutted by sufficiently strong experimental evidence, but if so, we should be driven to posit some radically new physical entity—as once electric and magnetic fields were posited—to account for the phenomena. Even more deep seated is our belief that the experimenter's interventions are not covertly correlated with some unknown factor. If in a run of experiments I repeatedly alter one factor, I do not believe that there are other, masking factors which are also altering, as it happens, just in step with my arbitrary alterations.[17] If I am right in this belief, then I can say that the experimental situations are indeed the same in all other relevant respects, and can attribute any difference in the effect to my intervention, either the factor I altered or some concomitant of it. The fourth presumption—that of the relative unimportance of the remote—could also prove false. There is a correlation, I believe, between sunspots and the trade cycle in China. If it continued to hold in spite of all sorts of changes in the Chinese economy, we might eventually be forced to allow that there was some causal connexion. But if it were to be an intelligible connexion, there would have to be some intervening link—say that the sunspots affected the weather, and the weather affected the trade, and it would be the weather we should identify as a more local cause. If it were an occult cause, it would be difficult to see why it was so specific. Why just China? Why not Norway, San Marino or the Argentine? Of course, it is logically possible that we should just have to accept it as a brute empirical fact, that the trade cycle of China depended on the sunspots, and that of the Argentine on the relative positions of Mercury and Mars. Astrology is not a logically impossible discipline. But since there are no rational grounds for picking out China as the sunspots' sphere of influence, we should need exclusively empirical grounds. If they are forthcoming, sunspots cease to be an unknown factor, and when we want to determine the effect of altering another factor, we check that the sunspots are the same in both cases. What we discount, on the strength of our fourth presumption of irrelevance, is the possibility, unsupported by any actual evidence,

[17] See above, Ch. III, p. 36.

that some unknown factor is making a difference to two apparently similar cases. In an old-fashioned terminology, we are rejecting the doctrine of Internal Relations, so far as causality is concerned.

These four presumptions entitle us, in default of contrary indications, to regard suitable cases known to differ in certain respects as being the same in all other relevant respects, and so enable us to make some causal judgements with rational confidence. Once we have discovered some causal connexions, we can begin to establish other, less profound presumptions of causal irrelevance. We learn by experience that it does not usually matter what colour the wallpaper is, whether the lab is heated by gas or by electricity, or how it is lit. We can be more confident when we know the sort of phenomena being investigated that whole ranges of factors are irrelevant. The earth's magnetic field is seldom relevant to chemical experiments, nor if we are measuring the frequency of sodium light. These and many other irrelevances are available when we seek to discover a particular causal connexion. Scientific enquiries are not conducted *de novo*. Characteristically, although not absolutely necessarily, we stand on other men's shoulders, and once some scientists have been able to stand on firm rational ground, we can avail ourselves of contingent discoveries already made, when we in our turn come to look for the cause of a particular phenomenon.

Granted some canons of irrelevance, we can test causal hypotheses, following exactly the same procedure as we did earlier when we knew exactly what the possible causal factors were. The procedure is essentially Mill's Method of Difference. As before, it enables us to pare away in a suspected causal situation the inessential circumstances, leaving only the kernel of types of which an event cannot be without the effect following. We have been able to do this by keeping "all other respects" *either* constant *or* irrelevant throughout our series of observations or experiments. We have therefore not ruled out the possibility: that it is only in combination with some of these other, relevant but constant, respects, that A_2 & A_3 is a sufficient condition of Z. Indeed, in almost all cases, causes operate only against a background of normalcy. Cigars and port will not make me ill if I am killed in a motor accident first. If the sun were to explode or the earth to collide with a new planet, few scientific experiments would yield their accustomed results. It is an essential condition of normal causes operating normally that the sun should not explode nor the earth collide. But it would be pointless to spell this out every time. Instead of specifying causally relevant but constant factors, we take them for granted as background conditions, knowing that since they do not alter, they will not make any difference to the different experiments we carry out. If any did alter, we could take them into account as potentially relevant factors. The boundary between specified causal factors and background conditions is an arbitrary one, and we can always specify a background condition as a possible causal factor. Although we cannot test *every* possible relevant combination of factors, we can test *any* particular combination that anybody cares to specify. So that if there is any reason for serious doubt, we can put the matter to the test, and settle it one way or the other. But there may, of course, still be other relevant factors, not obviously being taken for granted, which may alter without our knowing it. It is against this possibility that induction by simple enumeration and Mill's Method of Agreement give help. Each new instance of the regularity, differing from other instances as it must, at least slightly and in some condition or other, helps, if only by a little, to provide evidence that some further difference is irrelevant. And the number of possibly relevant differences is believed not to be inexhaustible. Induction by simple enumeration is a respectable canon of inference, provided that the instances are drawn from a wide range of circumstances. It is really a method of addition for proving causal irrelevance in the presence of a cause, not, like Mill's Method of Difference, a method of elimination in the absence of a cause. So too is Mill's Method of Agreement. If a cause operates under very different conditions, it is a fair guess that none of the background factors were causally relevant. It is a fair guess, not a conclusive argument, because we have, and, as we have seen,[18] can have, no criterion of being altogether different. If on one occasion when A_2 & A_3 were followed by Z the background conditions included B_1 and B_2, and on another very different one they included $\overline{B_1}$ & $\overline{B_2}$, it might still be the case that $(B_1$ & B_2 v $\overline{B_1}$ & $\overline{B_2})$ was an essential ingredient of the cause, and that A_2 & A_3 would not be followed by Z against background conditions which included B_1 and $\overline{B_2}$. In such an eventuality we do not abandon our claim that A_2 & A_3 is the cause of Z, but rather specify the conditions—

$(B_1$ & B_2 v $\overline{B_1}$ & $\overline{B_2})$

—under which it is, indeed, the cause of Z; or more commonly

[18] Ch. II, p. 26.

identify the conditions under which it is not the cause of Z—B_1 & $\overline{B_2}$ and $\overline{B_1}$ & B_2—as inhibiting factors. Instead of stating the minimum sufficient condition of Z, we state only part of it—a sub-minimum condition—and then state separately as negative conditions those under which the cause fails to operate. The workaday concept of cause thus begins to acquire an "open texture": Hume's constant conjunction does not have to be that constant; provided striking a match is normally followed by its lighting, we say that it is the cause, and explain away apparent exceptions as due to damp, absence of oxygen, etc. Instead of the cut-and-dried logic of necessary and sufficient conditions, we find ourselves operating with the looser, although more convenient, system of claim, other things being equal, and exception, when they are not and exceptional circumstances obtain.

We have little difficulty in operating with the everyday concept of cause or checking causal hypotheses in practice, but theoretically our procedures are less than watertight, and this gives an opening for the sceptic. We can be led to erroneous conclusions in inductive argument through our ignorance of the unknown. An unknown background or negative condition may change between different occasions of testing hypotheses, and yield anomalous results. Perhaps I eliminated oysters wrongly as a cause of my being ill, since on the two occasions, in the preceding and following terms, when I ate them without disaster, there was an R in the month. Such errors occur. But they are soon discovered. And often, even when we are dealing with unknown factors, we know enough, thanks to our previously discovered causal laws and scientific theories, to know what sort of factors to look out for as being potential background or negative conditions. So far as the practicalities go, we make mistakes, but not for long; and many causal hypotheses are established or refuted beyond reasonable doubt or the serious possibility of there ever being subsequent evidence to force on us a substantial change of mind. Theoretically, however, we may feel uneasy. It is not just that inductive arguments are not deductive, but rather that we can see how the unknown factors, which stand in the wings of every inductive argument, might conspire to mislead us, and how we might mistake for a causal connexion some concomitance which was not causal at all but due to other factors or to a Cartesian deceiver or a pre-established harmony or was simply a coincidence.

An apparent causal connexion may be due to some other systematic arrangement of factors. The classical example is that of the two clocks, say St Mary's and the Cathedral, each keeping perfect time, and the chimes of the Cathedral always following those of St Mary's after a short and constant interval. We might then suspect there to be a causal connexion, and an examination under a wide variety of different conditions would eliminate many other possible causes, but never this one. Yet we would be wrong to say the chiming of the one clock was the cause of the striking of the other. To eliminate this hypothesis we need to use the third canon of irrelevance to introduce a completely random factor into the situation. We interfere with nature and disrupt any pre-established harmony; regularities that survive our intervention can then be safely accepted as necessities and not merely engineered or coincidental contingent concomitances. We test the two clocks by moving the hands of St Mary's so that it strikes before the appointed hour; if the Cathedral strikes at the normal time we have proved that it is independent of St Mary's; but if it also strikes early, we conclude that it was a master clock in St Mary's we were tampering with. We argue that in general the effects of our actions are known, and there is no reason to suppose that our actions at St Mary's should, independently of the clocks being causally connected, bring about the striking of the Cathedral clock. If the concomitance had not been due to a direct causal connexion but merely to a common cause, our bringing about the one event by means of some other cause would not have brought about the other event and we should have obtained a counter-example to the concomitance. If we do not get a counter-example, then the concomitance was not due to a mere common cause or a contingent concomitance, but a genuine causal connexion. For our intervention was random with regard to other circumstances, which would therefore be in general different from what they had been in the previous, observed, cases of the Cathedral clock striking shortly after St Mary's had struck, and otherwise being silent. Hence a few experiments rapidly convince us that all the other circumstances are either unaltered from what they were when the Cathedral did not strike or else irrelevant, and using elimination by addition, we are left with the St Mary's clock as the cause of the striking of the Cathedral clock.

Our argument depends on two assumptions: a low-level, pedestrian one that our intervention did not have some side-effect which affected the Cathedral clock independently of our altering St Mary's; and the high-level presumption of irrelevance of states of mind *per se*. The former assumption, using the fact that the effects of our intervention are known and eliminable, is a contingent assumption which can be false. In biological experiments, notably, we cannot assume that our intervention is irrelevant. If we wonder whether it is flies that cause distress to cows, or whether it is some common factor that causes them both, say the hot weather, we cannot make a laboratory experiment of applying flies artificially, for the laboratory itself is enough to cause cows distress. In physics, however, we can often be reasonably confident that the effects of our intervention could not have had any covert effect on the second clock. The latter assumption is also contingent. It is conceivable that there should be a Cartesian demon who knows all our thoughts, and whenever we make up our minds to introduce an arbitrary alteration into the world, rearranges the course of events to give the appearance of a causal connexion. It is also conceivable that determinism is true, and that our making up our minds to intervene is part of the causal process and not really random with regard to other causal processes. Although often the processes leading up to an experimenter's deciding to make an arbitrary alteration are, as it happens, unconnected with the causal processes being investigated, it could be that there was a common cause which both led me to alter what I thought was the master clock in St. Mary's and caused the Cathedral clock to strike differently. Or, again, there is no inconsistency in attributing the concomitance neither to a Cartesian demon nor to a determinist pre-established harmony, but simply to coincidence.

The Cartesian demon cannot be confuted on its own terms. It is essentially an exercise in logical possibility. There is no logical impossibility, no inconsistency, in describing a world in which our every action is met by a response contrived to mislead us. But that is only to say that our resources of language are very wide. All sorts of things are describable. But the fact that something can be described without inconsistency is no reason to think it true. Before we think it true, before we even think that it might be true, we need to have some reason beyond bare describability for supposing it might be true. And no reason has been given. If the Cartesian demon is put forward as a serious reality, it needs to be substantiated, or else it is an idle suggestion without substance: if it is mentioned as a bare possibility, it is merely making the point, in a colourful way, that many of our canons and presumptions of scientific reasoning are not analytic. Which we knew already.

The thesis of a determinist pre-established harmony is equally implausible, but less easy to dismiss. For the most part human interventions, like natural disturbances, are unlikely to mesh with a particular causal process, and so are likely to disrupt any chance concomitances. But we attach greater significance to experiment than mere observation, partly because we have a stronger sense of randomness in our own arbitrary interventions than in the interplay of naturally occurring factors. It is against the absolute non-necessity of the experimenter's intervening that the necessity of causal connexion stands out. We see the experimenter as an absolutely independent outside agent, whose interventions could not be linked by some common cause to the physical process being investigated.[19] We could be wrong. But if we were, we should no longer be entitled to understand causality in the way we do or have the confidence we do that concomitances which survive random interventions must be genuine cases of cause and effect.

The third case to consider is that of coincidence. It might be that two quite independent processes were going on, and we were getting constant concomitance for no reason except the chance fact that the two processes happened to keep in step. If an arbitrary disturbance in the one is followed by a corresponding alteration in the other, it always could be that it was a genuine coincidence, that the latter had been going to happen anyhow, that if only we had not been so impatient we should have observed the processes getting out of step, and that it was our intervention that put them back in step again. But to argue this persistently is to make the same illicit extension of 'coincidence' as some phenomenalists do of 'illusion'. Coincidences, like illusions, are essentially things that do not much happen: we mean by them things that will not be repeated, runs that will not last. Here we are already extending it slightly to cover the case of two regularities that have got contingently in step. But to extend it further, so that the one regularity is in step not with another's regularity but with an experimenter's arbitrary and random interventions, is to stretch it too far. It is no longer a practical possibility

---

[19] Compare P. H. Eberhard, "Bell's Theorem without Hidden Variables", *Nuovo Cimento*, II, 38, 1977, B, p. 77. "It would not make sense to talk about non-actual experiments if, for instance, we believed that there is no free will, and the choice of the experimenter is determined anyway."

we are needing to eliminate but a Cartesian doubt. We need arguments, not further tests, to cure scepticism, just as we need arguments, not further sense experience, to cure phenomenalism.

The logic of cause and effect is a logic of minimum sufficient conditions, which in turn stems from the requirement of repeatability, or, in Hume's terminology, constant conjunction. The logic is complicated by our differing concerns. Our practical concerns, which we have only touched upon, lead us not only to require that the cause be weakly antecedent to the effect, but to pick out from a minimum sufficient condition as the cause *par excellence* only those factors that are variable or are easily manipulated by us or are of lasting significance or would justify an ascription of responsibility. Our theoretical concerns lead us to develop our concept of cause in a different direction. Already Hume, in his fourth rule for judging causes and effects, stipulates that there shall be a one–one relation between cause and effect, and that the cause should be not only a sufficient, but also a necessary condition of the effect.[20] The causal relation then becomes a mapping in a "logical space" of causal situations onto effect situations. We shall take this up in Chapter X. But first we need to elucidate the relations between the concept of causality and those of space and time, and in particular the apparent contradiction between the implicit commitment, in the requirement of repeatability, to the irrelevance of spatial and temporal differences *per se*, and the equal commitment, in "contiuity", to their relevance. In the next chapter we examine the concepts of contiguity and continuity, and why scientists adopt the postulate of "contiuity", and what follows from it. Physics in general has moved from discrete qualitative concepts to continuous quantitative ones, and in Chapter VI we consider measurement and how we are able to assign numerical values to physical features. We then return to repeatability and see how the requirement of universality, which underlies it, becomes in a numerical context the much deeper principle of invariance on which is based the fundamental conservation laws of physics.

[20] See above p. 44.

Further Reading

B   G. H. von Wright, *A Treatise on Induction and Probability*, London, 1951 or paperback edn., Paterson, NJ, 1960, Ch. IV.

B   J. L. Mackie, *The Cement of the Universe*, London, 1980, esp. Ch. 3.

For a different approach see

B   Donald Davidson, "Causal Relations", *The Journal of Philosophy*, LXIV, 1967, pp. 691–703.

Preliminary Reading for Chapter V

A   Wesley C. Salmon, *Space, Time and Motion*, Dickenson, Encino, Calif., 1975, Ch. II, pp. 31–52, 62–6.

B   Hume, *Treatise*, Bk. I, Part II, Sect. I–V, pp. 26–65. (See also *Enquiry*, Sect. XII, Part II, §§124–5.)

V

"Contiuity" and the *A Priori*

H UME'S empiricist critique of causality forced him to regard the causal relation not as being something in the external world and discovered by us, but rather as being something imposed on our understanding of the external world by our minds in consequence of their being conditioned to do so by the association of ideas. In this last conclusion he was wrong. Kant agreed with him that causal connexions were not so much discovered by our minds in the world as imposed by them on the world, but regarded the principle of causality as a regulative principle of reason, to be justified by argument, and not a mere habit, to be explained but not justified, as a form of psychological conditioning. The principle of causality is a necessary condition of our being able to make sense of experience, and is thus a schema we use to organize our sense-data into a coherent conceptual form. It is thus in an important sense *a priori*, although not analytic; and much of Kant's philosophizing is concerned with how such synthetic *a priori* propositions are possible. Both Hume and Kant agree at least in regarding causal relations as being not discovered in the world but imposed on it.

Hume, as we saw in Chapter III, was led to this conclusion because he regarded causal relations as being unlike spatial and temporal relations. These latter, he assumed, are perceived directly in experience, and many empiricists have agreed with him. They hold that distances and durations are immediately given in experience, and that all our theories of space and time are built up on these spatial or temporal sense-data, in much the same way as our theory of colours or of sounds is built up on what we see or hear. But this is wrong. There is a strong *a priori* element in our concepts of space and time, just as there is in our concept of cause. Although it may have been because causal relations are unlike spatial or temporal relations that Hume was led to regard causality as not wholly empirical, space and time are in this respect like causality, and contain a large *a priori* element too.

We can see that there is some *a priori* element in our understanding of time and space if we consider the presumptions of irrelevance we make in operating our concept of cause, and sometimes in justifying induction. We assume that a mere difference of temporal date or spatial position is, like a difference of person or subjective attitude, *per se* irrelevant. If A really is the cause of Z, then it must be so not only on Wednesdays but on Friday 13ths as well, and not only in Oxford, but in Cambridge, and not only when a lucky spell has been silently said before the experiment but equally when the experiment was conducted in the absence of any such secret incantations. These assumptions are synthetic. They rule out certain results, and we can imagine the empirical evidence being such that we were forced to abandon them. Indeed, physicists have recently been forced to abandon the fundamentally similar principle of the conservation of parity as a result of the experimental observations of the disintegration of the cobalt-60 nucleus. Nevertheless, although synthetic, they are not in any ordinary sense *a posteriori*. We assume them in advance of evidence, rather than establish them on the basis of evidence. No doubt, if the evidence were conclusively unfavourable, we should school ourselves not to make rash assumptions. But that we do in fact make such assumptions is shown by our rational reluctance to abandon them in the face of *prima facie* counter-evidence. We discount findings in favour of psychokinesis. If we find a causal correlation not being repeated at a later date we look for some further factor differentiating the situations in which it does not obtain from those in which it does. If a difference of spatial position or orientation appears to make a difference to the outcome of an experiment we posit some condition itself not uniformly distributed in space rather than attribute the difference just to space itself. The tendency of magnetic needles all to align themselves in the same direction is taken to show the existence of a magnetic field, not that space itself has a preference for the N–S axis.

That these assumptions are made can scarcely be denied. They are to be defended not on the grounds that they can be sufficiently established by empirical evidence but because they are necessary assumptions if we are to be able to argue inductively or apply the concepts of cause and effect. If a difference of date was by itself a relevant difference, no experiment would be repeatable: if a difference of position made a situation different, we could never have a constant conjunction of cause and effect, because different causes in different places would be therefore different, and not apt to produce the same effect.[1] Since situations that are numerically distinct must differ in temporal or spatial position, numerically distinct positions can be qualitatively identical only if temporal or spatial differences do not count as causally relevant differences at all. The principle of causality thus presupposes that of the intrinsic irrelevance of differences of spatial or temporal position, and hence our ideas of space and time have in them at least this *a priori* element.

Our concepts of space and time show other *a priori* features too. Space and time are taken to be homogeneous, and subject to further sophisticated symmetry conditions. Our criteria of identity and individuation are that a thing can only be in one place at once and can only move along a "contiuous" path, and, as we saw in Chapter III, although we hesitate to affirm it as an absolute principle, we are inclined to regard as perfectly acceptable and unproblematic only those causal connexions that are spatio-temporally "contiuous". If we take these requirements as being ones of strict contiNuity, then further topological questions arise. The dimension number of a continuous space is a topological invariant: whether the fact that space has three dimensions is an *a priori* truth may be questioned; but certainly the fact that time does not have more than one dimension seems to be more than a mere empirical truth. And questions of whether space is simply connected and orientable and whether time is cyclic or not are not in any ordinary sense left open to be settled by experiment.

Our concepts of time and space are in one sense inherently relational. Although particular instants and particular places may have their own individual names—my birthday, the dawn of creation, the first Easter, home, Oxford, Athens, Jerusalem, the centre of the universe—and these names may be expressed in entirely non-relational ways, yet it is always possible, and necessarily so, that we can raise the question with any temporal instant whether it is before

---

[1] See, e.g., E. P. Wigner, *Symmetries and Reflections*, Bloomington, Indiana, 1967, p. 4. "It is . . . essential that, given the essential initial conditions, the result will be the same no matter where and when we realize these. . . . If it were not [so], . . . it might have been impossible for us to discover laws of nature."

or after some other temporal instant, and, although perhaps less insistently, with any spatial position how it is related to other spatial positions. Although we can at least understand the concept of disconnected spaces—that is, spaces in which there are pairs of points in no spatial relation to each other—we would not call it a space if every point was in no relation to any other. Such a set would be regarded as being only a set, and as not having enough structure to be described as a space. We might go further, and say that a space is entirely characterized by its relational structure, and that no point can have any significant properties of its own, but only by reference to other points. Consideration of this relationist thesis we defer until Chapters VIII and IX: for the present we maintain only that space and time necessarily have some relational structure.

We consider time first, because it is simpler. Temporal relations form an ordering. If x is before y and y is before z, then x is before z. That is, temporal priority or antecedence is TRANSITIVE. Moreover, if x is before y, then y is not before x. That is, temporal priority is ASYMMETRIC. It follows that x cannot be before itself—that temporal priority is IRREFLEXIVE. Similarly for x being after y, etc. These three features, transitivity, asymmetry and irreflexivity characterize those relations that are ORDERING relations. They should be compared and contrasted with EQUIVALENCE relations, which are transitive, symmetric and reflexive. The paradigm examples of ordering relations are those expressed by the comparatives—taller than, larger than, slower than. In mathematics we often express an ordering relation by $>$ or $<$. We should note that we could also approach the concept of an ordering relation through the weaker relations $\geq$ and $\leq$ (e.g. taller than or the same height as), which are transitive, reflexive and antisymmetric (i.e. if $x \geq y$ and $y \geq x$, then $x = y$). This often has mathematical advantages but we shall not consider it here.[2] What we do need to consider are certain further questions that can arise about orderings. Normally we assume that temporal antecedence is "connected". That is, not only can we ask of any two instants whether one is before the other, but that always we can get

---

[2] The reader unfamiliar with the logic of relations should consult Patrick Suppes, *Introduction to Logic*, Princeton, 1957, §§10.5 and 10.4; or E. J. Lemmon, *Beginning Logic*, London, 1965, Ch. 4, §5; or R. R. Stoll, *Set Theory and Logic*, San Francisco, 1961, §11, pp. 48–52 (or, equivalently, *Sets, Logic and Axiomatic Theories*, San Francisco, 1961, §1.10, pp. 47–55); or H. B. Enderton, *Elements of Set Theory*, New York, 1977, pp. 62–4, 167–71; or W. H. Newton-Smith, *The Structure of Time*, London, 1980, Appendix, pp. 243–5.

a definite answer, either that x is before y or that x and y are the same or that y is before x. This assumption does not hold universally—not, for example, in the Special Theory of Relativity. There we have, for distant events, not a simple but only a partial ordering, in which one event cannot be unambiguously said to occur definitely before the other or simultaneous with it or after it. For "local time", however, such as we use in our ordinary life, we do have a strict ordering. That is, we can always say of temporal instants either that the one is before the other or that the one is after the other or that they are both the same.

A further question we need to raise about an ordering relation is whether it is discrete or dense, and if it is dense whether it is "gappy" or has the same sort of ordering as the continuum. Discrete orderings are those in which the word 'next' applies. A member has a next member, and there is no further member of the ordering between those two. Thus, if I am arranging towns in order of population, two towns cannot differ by less than one inhabitant, and it makes perfect sense to talk of the next larger or next smaller town. If, however, we were arranging them in order of area we should be hard put to it to say what the minimum difference was. Is it an acre, a square yard, a square inch, a square millimetre, a square Ångström unit? Although we should not think it worth making excessively fine distinctions, and could not make infinitely fine distinctions even if we wanted to, there is no lower limit to the fineness of distinction we might have occasion to make and might be able to make, with sufficiently refined methods. Between any two members, such that $A > B$, it is always possible that we might find ourselves wishing to place a third, C, between them according to this order, and say that $A > C$ and $C > B$. This leads us to say that the ordering is DENSE. In such cases it is appropriate to have our mapping into the rational numbers rather than the natural numbers or integers. The rational numbers—the fractions—have the property of being dense: between any two rational numbers, say 10/11 and 111/121, there is always a third, for example $(10 + 111)/(11 + 121)$ which reduces to 121/132 $= 11/12$.

Mathematicians use small Greek letters to name different sorts of ordering, or order-types as they are sometimes called.[3] The paradigm example of a discrete ordering is that of the natural numbers, 1, 2,

[3] For a full account of order-types, see E. V. Huntington, *The Continuum*, Dover, 1955, Chs. III, IV and V.

3, . . . etc. This is called the order-type ω. What is the order-type of the non-negative integers, 0, 1, 2, 3, . . .? It is the same. We might express this by writing $1 + \omega = \omega$. The two orderings are isomorphic because the latter can be mapped one–one onto the former, by the Successor relation. It will not work for an ordering on a finite class. Dedekind makes this, in effect, a definition of an infinite class. We should note that ω lacks a certain sort of symmetry. It has a first member, but no last one. Would it be possible to have a discrete ordering with a last member but no first one? What about the mirror image of ω, which would be . . . 3, 2, 1? This we call ω^*. Its natural example is the negative integers, . . . -3, -2, -1. If we adjoin the negative integers to the non-negative integers (*not* the natural numbers, or we shall miss out 0), we have an ordering of all the integers

$$\ldots -3, -2, -1, 0, 1, 2, 3 \ldots$$

We call this $\omega^* + \omega$.

Kant argues in his First Antinomy of Pure Reason that ω^* could not be a possible order-type of time, because "the infinity of a series consists in the fact that it can never be completed through successive synthesis". He goes on, "It thus follows that it is impossible for an infinite world-series to have passed away, and that a beginning of the world is therefore a necessary condition of the world's existence."[4] With the aid of the concept of an order-type, we can distinguish the logical from the temporal sense of 'succession', and see how ω^* can be infinite even though no member of it has more than a finite number of successors. We can also put forward the order-type $\omega^* + \omega$ as one which has no first member, but also no last one, and thus satisfies Kant's requirement while escaping his criticisms.

The paradigm example of a dense ordering is the rational numbers, ordered by size. Their order-type is called η. We should note that under another ordering the order-type may be different. Thus when we prove that there are only a denumerable infinity of rational numbers, we arrange them in a 2-dimensional array, and weave our way through them diagonally, and thus show that under *that* ordering they have the order-type ω.

Exercise

(1) Express in the first-order predicate calculus with Identity that a

---

[4] *Critique of Pure Reason*, A 426/B 454. Tr. Norman Kemp Smith, London, 1929, p. 397.

certain relation R is of order-type ω in the universe of discourse, or in some subset of it.[5]

(2) Express, similarly, that R is of order-type ω^*.

(3) Express, similarly, that R is of order-type $\omega^* + \omega$.

(4) Express that a is the R-est element.

(5) Express in the *second*-order predicate calculus that there is an ordering in the universe of discourse such that every set has a least member.

Hume regards space and time as discrete on epistemological grounds. There must be a minimum perceptible size and duration, and so, Hume argues, our ideas of space and time must be made up of these minimum perceptibles, and thus be discrete.[6] But these arguments again confuse conditions for the application of concepts with the meaning of the concepts themselves. Certainly, our discriminations of space and time are relatively crude, and therefore we can apply the concepts directly only to non-infinitesimal cases. But our discriminations can be refined by technical aids—e.g. photo-finishes—and there may be no limit to their fineness, nor to the smallness of intervals we may want to talk about or be able to assess by indirect methods. Hume's epistemological argument, therefore, lacks cogency. Although at any one time there will be a limit to our power of discrimination, so that, for all we know at that time, space and time may be discrete, there is not, in consequence, a limit to our power of discrimination which holds for all time. This is analogous to the argument that from every number's always having a greater it does not follow that there is a definite number that is greater than every other number; or in formal logic

$$(x)(\exists y)Rxy \not\vdash (\exists y)(x)Rxy.$$

This, of course, does not show that time is dense. Indeed, we clearly cannot show that directly. However finely we can distinguish instants or positions, it is always possible that there is an ultimate granular structure to time or space, beyond which we cannot discriminate further. Our beliefs about the structure of time and space are thus

---

[5] The last clause is to take care of cases such as $\omega + \omega^* + \omega$. It is impossible in the *First*-order Predicate Calculus to exclude all such cases.

[6] *Treatise*, Book I, Part II, Sect. I and elsewhere.

beliefs which cannot be directly verified, although they could perhaps be falsified. Nevertheless, we do regard time and space as being not discrete, but dense. This should be seen not as a simple empirical fact but a conceptual truth. It is discovered not by experiment but by reflection. It is not that we have observed time not being discrete, but that when we think about it, we realise that we think of its being dense—that is, although we sometimes talk about "the next moment", there always is, as indeed there has to be, an intervening interval in which other moments lie.

Contrary to what we might at first suppose, a dense ordering can still have gaps. This was discovered by the Pythagoreans, who proved that $\sqrt{2}$ was not a rational number: that is, there is a gap in the rational numbers, all of which are either such that their squares are greater than 2 or such that their squares are less than 2, but no one of which is such that its square is exactly 2. This was for over 2,000 years a great weakness in the foundations of mathematics, and was finally made good by Dedekind and Cantor. The continuum of real numbers has no gaps. It is much, much more fully filled than the rational numbers. We could not possibly rearrange the real numbers to have the same order-type as the rational numbers (in the way the rational numbers could be rearranged to have the same order-type as the natural numbers), because there are, as we know, $2^{\aleph_0}$ real numbers, but only $\aleph_0$ rational numbers, where $\aleph_0$ (pronounced Aleph nought) is the cardinal number of natural numbers, $1, 2, 3, \ldots$, etc.

The question therefore arises whether we should ascribe to time the order-type of the rational numbers, η, or the order-type of the real numbers, θ, as it is sometimes called, or more frequently in modern texts, $\mathbb{R}$. Many physicists are impatient with the question, reckoning that it has no empirical content. But although it is a conceptual question, it is not an idle one. It cannot be immediately verified or falsified, but it does have important consequences for the whole way we understand physics, and in analogous cases can yield consequences that may be tested by experiment. Apart from the direct argument, that a "gappy" time is highly counter-intuitive,[7] the most accessible *a priori* consideration for our needing to conceive of time and space as continuous is that we have occasion to use, e.g. quadratic, equations which have solutions in the real numbers, but not among the rationals. If we have the distance–time law

[7] J. R. Lucas, *A Treatise on Time and Space*, London 1973, §6, esp. pp. 33–4.

$$s = ut + \tfrac{1}{2}at^2$$

we shall have temporal intervals whose ratio will be $\sqrt{2}$; or, to put it another way, if there is a point, some 16 feet below, where a body falling under gravity will be exactly one second after being dropped, then there is no instant at which it has reached the half-way point. So, too, if we have more than one dimension of Euclidean space, we shall, as the Pythagoreans realised, have lines whose length is $\sqrt{2}$ times that of another line. In a similar way, we want to say that the altitude of an equilateral triangle is $\sqrt{3}/2$ as long as the sides, the circumference of a circle $\pi \times$ diameter and the period of a pendulum $2\pi\sqrt{(l/g)}$. Each of these leads to irrational numbers, and it is a natural assumption for the mathematician to suppose that to each real number there is a corresponding interval, point or instant.

Again, since there are no gaps in a continuum, we can apply analogues to the argument by "Mathematical Induction". Mathematical Induction (to be sharply distinguished from the induction discussed in Chapter 2) applies only to discrete orderings in which there is always a *next* number, and says that if something obtains at the outset and then always obtains for the next member, it obtains for them all. It is a principle of great importance in mathematics, but cannot be applied to dense orderings, where there is no next number. We can see the difference between the merely dense and a continuous ordering, if we consider a principle of universal causation stating that for every state of the world there is an earlier one which is its cause. Intuitively, such a principle is equivalent to the claim that any state of the world determines all subsequent ones, and so it is, *provided* time is continuous. But if time were merely dense we could satisfy the principle of universal causation without satisfying that of universal determinism. For example, we could have the state of the world at time t being caused by that at time $t-1$, and the state of the world at time $t-1$ being caused by that at time $t-1.4$, which in turn was caused by that at $t-1.41$, itself being caused by that at $t-1.414$, which again was caused by that at $t-1.4142$. We could go on, producing a whole series of earlier and earlier states of the world, converging to the time $t-\sqrt{2}$. But there would be no such instant, if time were merely dense, and we therefore could not apply the principle of universal causation to the state of the world at this instant in order to go back to a time before $t-\sqrt{2}$. At every instant after $t-\sqrt{2}$ the state of the world would have been caused by the state of the world at some time earlier, but none of them need have been caused by the state of the world at any time before $t - \sqrt{2}$. For all we can say, a new sequence of events might have started later than all instants before $t - \sqrt{2}$ but earlier than all instants after $t - \sqrt{2}$. Only if we can talk of the instant $t - \sqrt{2}$ and of the state of the world then, can we rule out this possibility, and argue, on a principle analogous to that of Mathematical Induction, from universal causation to universal determinism.[8]

A quite different argument for continuity arises from topology. Topology is concerned not with the metrical properties of spatial figures but those which would be preserved if space were elastic, and were stretched or compressed like a rubber sheet, but without tearing. The fundamental topological structure is a continuum—the topology of discrete or dense sets is degenerate. If space is continuous, it makes sense to talk of its topological properties in a simple, direct fashion. For example, dimension number is a topological property and invariant under any one-one continuity-preserving mapping, but not under mappings that do not preserve continuity. It seems natural to regard the fact that space has three dimensions as a fundamental fact about space. But if we are to regard dimensionality as a fundamental feature of space, we are committed to regarding space itself as a continuum.

A further, somewhat two-edged argument can be brought against the claim that space is discrete. If space had a fundamentally granular structure, it would have some preferred directions along the axes of packing, as in an apple orchard or a salt crystal. But space is, we believe, *isotropic*, that is to say, it has no preferred directions.[9] Hence it must be at least dense. This argument, however, only works for space, and not for space-time. In the Special Theory of Relativity there is a preferred speed, the analogue to a preferred direction, namely the speed of light, and it would be open to a physicist to argue—and has, indeed, been argued by a Russian physicist[10]—that this is due to a granular structure of space-time, in which light always goes at the next moment to the next point.

Other physicists have thought that quantum mechanics argues for a discrete structure of space and time. Most quantum theories,

---

[8] I owe this example to Professor G. H. von Wright. For an essentially similar characterization of continuity by Cocchiarella, see A. N. Prior, *Past, Present and Future*, Oxford, 1967, pp. 71–2.

[9] See below, Ch. VIII.

[10] A. N. Vyaltzev, *Diskretnoye Prostranstvo-Vremya*, Moscow, 1965.

however, are not committed to a discrete structure, but only to a minimum *unit*. There is a minimum unit of action which is always the same, and hence, under various conditions, minimum intervals of time and distance and minimum units of energy and momentum, that can occur in physical reality. But what is the minimum varies with the conditions, and is always *of* a magnitude, itself thought of as continuous.

These arguments do not by themselves rule out the possibility that there may be a fundamentally granular structure to space and time, though they circumscribe it. Although we do not at present have any discrete-space or discrete-time physics to make a comparison with, physicists could discover that there were quanta of distance or duration, as they have discovered the atomic nature of matter, the discrete structure of change and the quantum of action. Or they might be led to conclude that space and time were dense, but not continuous. Who can tell? None of these hypotheses can be ruled out as inconsistent; none of them is clearly incompatible with the evidence. Indeed, although there are considerations, some of them involving empirical evidence, which may *incline* us to adopt one hypothesis in preference to the others, there is no evidence which would *force* us to. We always *could* maintain that time was discrete, but that the units of time were very, very much smaller than anyone had supposed, or—the other way—that time was continuous, and its apparent granularity not fundamental. No empirical evidence could be absolutely conclusive. There is considerable philosophical significance in there being hypotheses logically incompatible with one another, each internally consistent and each compatible with all the empirical evidence. Quine speaks of theories being "under-determined by the data", and maintains that all theories are underdetermined and that in consequence there is an ineliminable indeterminacy in translation.[11] Newton-Smith believes that it seriously circumscribes the possibility of giving a realistic account of time.[12] Empiricist philosophers who believe that empirical evidence is the only touchstone of truth, are committed to saying that if there is no empirical difference between two theories, then there is no real difference at all between them, so that if they are logically different from each other, we must re-interpret them to slur over the difference. But that is uncalled for. There is no need to make out that

[11] W. V. Quine, "On the Reasons for the Indeterminacy of Translation", *Journal of Philosophy*, 1970, pp. 178–83.
[12] W. H. Newton-Smith, *The Structure of Time*, London, 1980.

different theories are the same if they are compatible with the same empirical evidence, because empirical evidence is not the only consideration we have to guide us. We may be guided also by the logicality and harmony instanced by Tolman as grounds for accepting the Special Theory of Relativity, or by the reasons given in this chapter for accepting the continuity of space and time.[13] These constitute rational grounds for preferring one theory to another. It is only if we adopt the tenet of radical empiricism that there are no rational grounds, other than inconsistency or incompatibility with the empirical evidence, for rejecting one theory as compared with another, that we have to treat underdetermination as a problem requiring desperate remedies.

The continuity of space and time is thus a test case as between radical empiricism on the one hand and various forms of rationalism and realism on the other. For the radical empiricist, it is an idle question whether space and time are continuous or not. Philosophers with some tincture of rationalism or realism, however, will not be driven to so counter-intuitive a position. Even though there is no empirical evidence to decide the matter, there may be other rational considerations bearing on it. Realists expect theories, and indeed most assertions, to *go beyond* the evidence on which they are based, and will not be in the least surprised at there being different theories going beyond the evidence in different ways. The theories differ in making different assertions about reality, and will be true or false according as to whether reality is or is not as they claim. There are, on this view, some facts of the case—space and time being continuous or space and time being merely dense—which make our statements about the continuity of space and time either true or false. It may be that we cannot see, and shall never be able to see, which of these cases obtains, but this does not prevent one or the other of them obtaining. God knows what the truth is. And although we cannot see directly what it is, we may be able to tell indirectly. Physicists develop their "physical intuition" which tells them which of several possibilities that are compatible with the empirical evidence is likely to be true; and we can hope to do the same.

Controversy rages over the merits of various forms of empiricism, rationalism and realism. Although somewhat counter-intuitive at first sight, radical versions of empiricism have been in the ascendant over much of this century, partly as a result of Einstein's empiricist

[13] See above, Ch. I, p. 5.

critique of the concept of simultaneity, partly as a result of the difficulties in giving any realist construal of quantum mechanics. Nevertheless there is a strong commitment to rationalism and realism in physics, and physicists are much guided by rational, as well as empirical, considerations, and instinctively committed to a belief in physical reality. These beliefs are usually not very clearly articulated. In particular, although simplicity, elegance and aesthetic beauty are often cited as marks of a theory's being true,[14] no adequate criterion of simplicity has ever been formulated. Moreover, as Newton-Smith argues,[15] we cannot be sure that of any two empirically equivalent theories one will be simpler than another, and therefore may always be faced with two different theories between which there is nothing to choose. But this is only a 'may'. Often in the history of physics there have been rival theories, both of which were equally supported by the evidence and neither of which appeared at the time to be simpler or more elegant than the other. But standards change. The immensely complicated corrections to mass and spatial and temporal intervals required by the Special Theory of Relativity in the early part of this century now seem a simple corollary of our use of 4-vectors. Although at any one time there may be competing theories in the field, in the fullness of time we refine our standards, and become able to tell which seems more rational and more likely to represent physical reality.

In order to establish radical empiricism firmly we should need to have a clear criterion of the range of possible empirical evidence and to discount all putatively rational considerations of simplicity, harmony, logicality and elegance. Neither of these conditions is satisfied. Physicists are constantly devising new ways to put theories to an empirical test; and they are guided by a large variety of considerations beyond those of logical consistency and compatibility with empirical evidence. Although the alternatives to radical empiricism are not clearly spelt out, and are as yet unsatisfactorily messy, they seem closer to the truth.

We have taken the basic topological property of continuity as a test case between radical empiricism and various forms of rationalism and realism. Other topological properties are equally apt. Kneale offers, as his prime example of a metaphysical statement

[14] e.g. by R. G. Swinburne, *Space and Time*, London, 1969, p. 51; compare Richard C. Tolman quoted Ch. I, p. 5 above.
[15] W. H. Newton-Smith, *The Structure of Time*, pp. 72-3.

which is neither nonsensical nor merely verbal, the thesis that time is not cyclical, but Newton-Smith maintains that the theses that time is, and that it is not, cyclical, are further examples of theories underdetermined by data, so that once again there is no real point at issue between them. If we believe it is a real question, we are realists: if we believe that not being conclusively decidable, it is a non-question, we are radical empiricists. Other typically topological properties of time and space are that time had a beginning or will have an end, that space is unbounded and simply connected, that time has a direction, that space is orientable, that time has only one dimension, and that space has three.

The continuity of space and time thus turns on fundamental principles of philosophy and gives rise to a number of sophisticated and significant issues in their topology. Many other of our concepts will be much more complicated than Hume supposed. Instead of simple inspection of definite properties of discrete objects, we shall have to operate procedures of measurement which will need careful justification. Instead of constancy we shall need the more refined concepts of covariance and symmetry, and instead of simple causal laws we shall need differential equations expressing functional dependences.

Further Reading

A   W. H. Newton-Smith, *The Structure of Time*, London, 1980, Ch. VI.

B   Edward V. Huntington, *The Continuum*, Dover, NY, 1955, Chs. III, IV and V.

[16] W. C. Kneale, "Time and Eternity in Theology", *Proceedings of the Aristotelian Society*, LXI, 1960-1, pp. 91-2: W. H. Newton-Smith, *The Structure of Time*, London, 1980, Ch. III.

Preliminary Reading for Chapter VI

A Wesley C. Salmon, *Space, Time and Motion*, Dickenson, Encino, Calif., 1979, Ch. II, pp. 52–62.

B N. R. Campbell, *Physics: The Elements*, Cambridge, 1920, republished as *Foundations of Science*, New York, 1957, Ch. X, pp. 267–94.

VI

Measurement

W E are very familiar with measuring: so much so that we assume that it is something very simple, and not in need of conceptual elucidation. But it is a complicated, "theory-loaded" operation, depending on many assumptions which need to be made explicit.

We tend to think of measurements as being exclusively concerned with numbers, but this is to narrow our concern unduly. Measurements are typically concerned with numbers, but not essentially so. Besides the standard cases where we assign a rational real number to a length, a duration, an angle, or a mass, there are some cases where we use not the rational numbers but the integers—Möhr's scale of hardness, atomic numbers—and some where we use a discrete two-valued scale—as when we assign positive or negative charge, or describe a magnetic pole as North or South. Generalising the other way, we should admit as cases of measurement those where we describe a field in terms not of scalar numbers but vectors or tensors. All these are measurements. In each case we assign to a material object, physical event or natural phenomenon some abstract object from a certain range of abstract objects, which may be discrete finite-valued—e.g. $(+, -)$, (N, S)—or discrete infinite-valued, or rational, or real—e.g. π radians—or complex, or . . . There are also cases where what is measured is itself something abstract, as when we impose a metric on an abstract space. Looking at it from a more formal point of view, we say: A SYSTEM OF MEASUREMENT IS A MAPPING OF MATERIAL OBJECTS, PHYSICAL EVENTS, NATURAL PHENOMENA, OR ABSTRACT OBJECTS INTO A RANGE OF ABSTRACT OBJECTS, such as numbers. We should note that this definition includes the assignment of co-ordinates and dates as well as more ordinary measurements. It could be argued that to assign co-ordinates to a material object or physical process or a natural phenomenon was exactly like assigning mass or duration. But we often assign co-ordinates and dates to positions, directions and instants irrespective of whether there are any material objects, physical processes or natural phenomena occupying them. We should construe these as cases where we assign numbers as a means of *referring to* positions, times and directions rather than *describing* any feature of the natural world. For example, if we give the latitude and longitude of a place on the earth's surface or a grid reference of a place in the British Isles or refer to a nor'easter, we are indicating what we are talking about rather than saying anything about it. Typically, and for good reason, our frames of reference are closely connected with measurements of a more normal kind. But we should distinguish them, regarding frames of reference as a special—perhaps even degenerate—case of a mapping into a range of abstract objects.

Why do we make measurements? Because we seek a certain sort of objectivity. Although on many occasions we are content with subjective descriptions, often, indeed, preferring them, because they are more vivid and illuminating, they are none the less liable to prejudice, error and bias, because they depend on the judgement of the person making them. If we are anxious to avoid the danger of being wrong, we go through a procedure whereby our answers will be checked, and made reasonably proof against error. It is this, rather than the use of a numerical vocabulary, which is essential to measurement. If I say I mind about the outcome of an election only half as much as I do about Oxford's losing the boat race, I have not made a measurement. Estimates are not measurements, although they are indubitably numerical; whereas a qualitative analysis in chemistry is a measurement, although it is not numerical. If I have been through the tests for a calcium salt and a chloride, then I can be reasonably certain that calcium chloride is present—far more so than if I had relied on my taste, or guessed from the colour. If I guess that the window is two yards wide, I am quite likely to be a few inches out, whereas with a modicum of competence, I shall be accurate to within an inch if I use a tape-measure. And this is important when I come to buy curtains or a curtain rail. Measurement, because it is objective, is not only intersubjective, but repeatable. If I measure the window, I can be reasonably sure that the curtain rail, when I come to put it up, will be long enough, and that if I order the curtains by post, the amount of material cut off by the draper will be no different from what would have been cut off if I had done it. Although on other occasions we may prefer less impersonal descriptions, and may describe the windows as spacious and the wine sauce as intriguing, if our concern is to minimise subjective error we shall fall back on procedures where a competent observer can scarcely be wrong. Although we cannot eliminate the observer altogether, we can reduce his role so as to minimise error. So we cut his role down to standard size. If it is difficult to be wrong about a meter reading, or a needle against a dial, or whether two weights are balancing against each other or not, and if we base our descriptions on operations such as these, we shall be reasonably immune from errors of judgement. Equally, if I give the latitude and longitude of my island in the South Pacific, or the grid reference of my farm in England, or my address as 4025, 46th St, NYC., or tell the subaltern to make the platoon advance to the N.E., people are much less likely to get me wrong than if I described the island, named the farm, gave verbal directions in New York, or talked to the sergeant about sunrise on midsummer's day.

Except when we are giving a reference, our measurements are a sort of description. In reporting the result of a measurement, we say that what is being measured has some sort of feature, property, quality or relation, which could in principle be possessed by some other thing if we were to measure that. Being 57 inches wide is not unique to my bedroom window alone—other windows, pieces of furniture, rolls of curtain fabric can also be 57 inches wide. In formal terms, the mapping into a set of abstract entities is not one-one, but, potentially at least, many-one; not an isomorphism, but a homomorphism. Any many-one mapping, any homomorphism, induces a partition into equivalence classes. We have the equivalence class of all the things that are 57 inches wide, the class of all the bodies that weigh 4 oz., the class of all the organisms that have blood temperature of 98.4 °F. This is a truism of mathematical logic. But in making measurements we assume that it is not a mere truism. When I measure my bedroom window, I not only go through a procedure which is such that any other honest and minimally competent observer will reach approximately the same result, but I want there to be some significance in assigning a width of 57 inches to the window *vis à vis* measurements of pieces of furniture and curtain fabrics. It would be no good if I and everyone agreed that the window was 57 inches, and that the furniture was 57 inches, and that the fabric was 57 inches, but there was nothing in common between the window, the furniture and the fabric other than the bare fact that we all assigned the description 57 inches wide to each. The whole point of describing them all as being 57 inches wide is that there should be something in common. Hence measuring does not *impose* but *presupposes* there being equivalence classes in the things that are measured.

The clearest example is weight. A pair of scales enables us to determine an equivalence relation. If two bodies balance against each other they are members of the same equivalence class. There is an appeal to spatial symmetry in our assumption that if the arms of the balance are equal, then the weights in the pans must be equal also, but it is subject to empirical check. We can make the experiment. I find that if two bodies balance against each other and we interchange them, they still balance against each other; that is, the relation of *balancing against* is symmetrical. It is also transitive. We find, within reasonable limits of experimental error, that if two bodies balance against a third, they balance against each other. *Balancing against* is thus an equivalence relation, and therefore can define a set of equivalence classes, each containing all those things which balance against each other and which will therefore characterize as having the same weight. If this were not so—and we could imagine a world in which it was not so—we could not form a concept of weight.

Exercise

Is *balancing against* an equivalence relation? Equivalence relations are reflexive, but a body cannot balance against itself. Formulate an accurate definition of *balancing against* in the first-order predicate calculus, and in terms of that an accurate definition of *being the same weight as*.

Balancing is a paradigm, and shows that measuring essentially involves equivalence classes. Although often, as when we use a spring balance, a ruler, or a dial, we seem to apply numbers direct, there would be no point in so doing unless other things besides the object measured could have the same number assigned to them, and unless in that event they had some other significant property in common. It is, first and foremost, the equivalence class which gives a measurement its significance. These equivalence classes constitute a partition of the set of measurable objects, and the purpose of measurement is to assign to each equivalence class an appropriate number which will be what each member of that equivalence class measures. We can then form the "Quotient Set"[1] of the set of measurable objects by the equivalence relation, and consider that as what is, abstractly, being measured. Instead of the potentially many-one mapping of measurable objects, X, in our definition, into a range of abstract objects, we have a one–one mapping of the quotient set $X/_\approx$, that is to say the set of equivalence classes of members of X, into the same range of abstract objects. Instead of the many–one mapping which correlates Peter with 12 stone 13 lb., Paul with 12 stone 13 lb., Patrick with 12 stone 13 lb. etc., we consider first the equivalence class of Peter, Paul, Patrick, *et al.*, who could be balanced against one another, and all weigh the same; and then we assign a number—12 stone 13 lb.—to the whole equivalence class. This may seem a finicky distinction, but it is useful both technically and conceptually. It is useful technically because, when we come to introduce order, it will enable us to work entirely with $<$, less than, a strict ordering, rather than with $\leqslant$. Weights and heights, that is, are strictly ordered, and provided x and y are distinct, either $x<y$ or $y>x$, whereas with people there is always the awkward chap, different from myself who is nevertheless exactly the same weight.

Balances not only tell us when two bodies weigh the same, but on those occasions on which they do not weigh the same, tell us also which of them is the heavier. That is, they give us an order. Ordering relations, as we have seen,[2] are a very important type of relation, as important as equivalence relations. Equivalence relations, we recall, are

TRANSITIVE   SYMMETRIC   and   REFLEXIVE

whereas typical ordering relations are

TRANSITIVE   ASYMMETRIC   and   IRREFLEXIVE.

Except in the case where we are mapping physical systems onto a discrete two-valued set—as when we determine whether a magnetic pole is N. or S.—the abstract entities we are using will have a natural

---

[1] See e.g. Robert R. Stoll, *Sets, Logic and Axiomatic Theories*, San Francisco, 1961, p. 34, (or, equivalently, *Set Theory and Logic*, San Francisco, 1961, p. 31); or Herbert B. Enderton, *Elements of Set Theory*, New York, 1977, pp. 55-61.

[2] Ch. V, p. 72.

order, which in almost all cases is simple (or connected) and strict (or linear).[3] Given any two different atomic numbers or wavelengths, either one is greater than the other or the other is greater than the one. In setting up a homomorphism, a many-one mapping, we need to find an empirical relation with the requisite ordinal properties, in the same way as we needed an equivalence relation holding between different things measuring the same. It is an empirical truth susceptible of experimental check, that if one body goes down when weighed against another, and up when weighed against the third, the third will go down when weighed against the other. And again, that if one body goes down when weighed against another, it will still go down if the bodies are interchanged in the scales. Equally it is confirmed by experiment that atoms of an inert gas with an anomalous liquid form at very low temperatures have two protons, whereas atoms of a gas produced when an acid reacts with a metal have only one, and that a beam of red light is bent less by a prism than a beam of yellow light, and a beam of yellow light less than one of blue. As with equivalence relations, it is an essential assumption of making measurements that the ordering relations which hold among the abstract entities correspond in some way to ordering relations holding among the things being measured. Even when we assign numbers merely to refer rather than to describe, we expect this condition to be fulfilled—the whole point of giving one's address as 4025, 46th St, rather than as Mon Repos, Acacia Avenue, is that the former mode of reference is systematic, and enables a stranger to find the house by a systematic procedure, whereas the latter, being *ad hoc*, depends on luck or extraneous knowledge. It is tempting to say that the ordering among the measurements should be exactly the same as that among the things being measured, but in fact there is a distinction due to the mapping being many-one. Weights and heights, lengths, angles and durations, and other abstract entities, are, as we have seen, quotient sets, and the ordering among them is not only simple but strict. That is, we order weights, heights, lengths, angles, durations and other abstract entities by the relation $>$, *greater than* (or, equivalently, by $<$, *less than*), whereas we should have to represent the ordering among things measured by $\geqslant$, meaning *heavier than or as heavy as*, or *at least as heavy as* (or, equivalently, by $\leqslant$, meaning *less heavy than or as heavy as*), etc. In the theoretical approach we have adopted, we stress the distinction, and

[3] For a full account of ordering relations, see works listed in Ch. 5, p. 72, n. 2.

then "factor" it out: we start by establishing equivalence classes, and, having done that, require only that physical systems from *different* equivalence classes shall be able to be simply ordered. In practice, however, we perform not two separate operations, but only one. We weigh one object against another, and find either that one is heavier than the other or that they both balance. In the former case we say that the weight of the one is greater than the weight of the other, in the latter that the weight of each is the same as that of the other. Thus essentially we reduce the ordering 'heavier than or balancing with' which orders material objects strictly but not simply, with the general order-type $\geqslant$, to a simple ordering of abstract entities, weights, which can be ordered simply by the relation 'greater than', with the general order-type $>$. This we can do because we are using a many–one mapping from physical systems to abstract entities.

From the logician's point of view, ordering relations can best be studied in the paradigm example of class-inclusion. If we consider classes or sets, and consider the relation of one set being included in another, we find that this is an ordering relation, and that all ordering relations can be represented by this relation within a suitable domain of sets. In natural philosophy the traditional formulation of the principle has been "The whole is greater than the part". We reckon that it stands to reason that two material objects together have a greater weight than either by itself, and that if B lies between A and C on a straight line, then the length AC is greater than the length AB, and that if event E occurs after D and before F, the interval DF is greater than the interval DE. For various reasons it is expedient to alter and sharpen the traditional formulation. In considering physical systems we want to use not the simple ordering relation 'greater than', $>$, but $\geqslant$, 'greater than or equal to', and correspondingly we need to make it explicit that we use class-inclusion in the sense in which every class includes itself, and would therefore write it $\supseteq$. The convention among logicians, however, is to work with the converse relation, of one class being included in another class, which is written $\subseteq$. The precise corresponding principle in natural philosophy would be "The part is less than or equal to the whole", understanding by 'part' not 'part' in its ordinary sense ('proper part' to the logicians), but in an extended sense in which the whole would count as being part of itself.

In many cases, whatever formulation we use, we have little difficulty in identifying some physical analogue of the part–whole relationship which will serve as a basis for ordering physical systems in respect of some physical property. But we need to make it explicit. Occasionally the ordering relation is not a part–whole one, as in Möhr's scale of hardness; and occasionally the part–whole relationship will fail to establish an ordering, as for example in the Special Theory of Relativity where a particle moving at the velocity of light in a frame of reference itself moving with a positive velocity in the same direction with respect to another frame of reference is *not* then moving faster with respect to the second frame of reference than the first frame of reference is. We shall consider this paradox more carefully when we come to consider not merely establishing an order but operating an addition rule for physical magnitudes.[4]

Once we have an order, we can raise the question of limits. Is there a least weight, a least duration, a least length? Or a greatest? Often it seems self-evident that there must—or alternatively that there cannot—be, but our intuitions may be at fault. Philosophers have held both that there self-evidently must be a first instant of time, and that there self-evidently cannot be. It would seem fairly obvious that there must be a least velocity—i.e. a state of absolute rest—and could not be a velocity faster than which it was impossible to go, but according to the Special Theory of Relativity, just the opposite is the case. If space is Euclidean, then there are no longest straight lines, but if space always had a positive curvature, then there would be "straight" lines, like great circles on the surface of a sphere, longer than which it would be impossible to be. It was far from obvious to our forefathers that there was an absolute zero in temperature but not an absolute maximum, and that things could not go on getting colder and colder, but could always be hotter without limit. Caution therefore is needed in laying down limits. We assume that there is a natural zero in weight, in duration, in distance and in angle, but the justification is not always the same, and not always indisputable. Negative weight was known. We have defined it away, by specifying that all weights should be measured in a vacuum, and theorized it away by deeming phlogiston not to be a substance. But negative mass is still conceptually possible. It is different with duration, distance and angle. Each of them is seen as a function of two terms—two instants, two positions, two intersecting lines—and the case where the two are coincident provides a conceptual limit in one direction. Duration, distance and angle have a conceptual zero; mass a natural one, but it is conceivable that our physics might lead

[4] See below, pp. 93, 100–2.

us to revise it; temperature a zero given in physical theory, but fairly remote from our familiar acquaintance with the concept of hot and cold.

Thus far we have established only an order, not a scale. We could measure degrees of weight, length, angle and duration, but not magnitude. That was for a long time the position over temperature, as the name still shows, and it is worth considering the assumptions involved in measuring temperature because they are relatively remote from the centre of physics, and hence more easily recognised for the assumptions they are. How do we normally tell how hot it is? We use a thermometer. But why should the length of a tube of mercury or alcohol have anything to do with how hot a thing is? In part it is a brute empirical fact. When the wind blows chill from the north-east in winter the thermometer reads something in the thirties, and on a sweltering summer day it registers in the high seventies. But this is not all there is to it. We are prepared, on occasion, to correct our subjective impressions of warmth by appeal to the thermometer, and the thermometer gives far finer discriminations than we can detect. We trust the thermometer not simply because of some crude correspondence with experience, but because we believe that mercury and alcohol expand as they get hotter. We could not have water in a thermometer, because water does not always expand with heat. Between 0 °C and 4 °C it contracts. We believe, and need to be able to believe, not only that mercury and alcohol always expand with heat, but that glass does not expand so much as to offset the expansion of the liquid. In practice we minimise the effect of the expansion of the glass by having a large bulb and a thin tube. But it could be otherwise, and we might have no transparent solid whose behaviour under change of temperature was discountable.

Granted these assumptions, we can use thermometers to indicate *degrees* of temperature; that is, the thermometer will reliably indicate whether one thing is hotter than another. But it still will not give us a *scale*. We have no warrant for thinking that one degree's difference is in any way the same when it is between 3 °C and 4 °C as when it is between 99 °C and 100 °C. It might be that the coefficient of expansion of mercury or alcohol was not constant. After all, we know that the coefficient of expansion of water is not constant. As it happens, water is an exception—a very fortunate exception—and for most liquids the coefficient of expansion, although different for different liquids, is more or less the same for any one liquid at all different temperatures. We could just form a "consensus scale", based on the assumption that for all non-exceptional liquids the coefficient of expansion was constant with respect to changes of temperature. But we would still feel dissatisfied with this. It is, fundamentally, a circular definition. We measure temperature by the expansion of liquids, and, having assumed that the coefficient of expansion is constant, discover that indeed it is so. We could parry the objection by pointing out that if the coefficient of expansion of liquids were not constant over temperature, we should expect that different liquids would have their coefficients not only different, but varying differently with respect to temperature. But still it could be the case that the coefficient of expansion of liquids itself increased uniformly with temperature—perhaps linearly, perhaps logarithmically or exponentially—or even that it was much greater below 49 °C than it was above it. We cannot really answer these doubts, so long as we are using only a consensus scale. We can fix the scale of temperature by *fiat*, but it remains an arbitrary convention, which could be out of accord with the real laws of physics.

The case is altered with the development of thermodynamics. If we believe in calories, and believe that one calorie will raise one gram of water through one degree centigrade, not only at 15 °C (the official definition) but at any temperature between 0 °C and 100 °C, we can check whether the heat given out by a gram of water as it cools from 100 °C to 99 °C will raise another gram of water from 0 °C to 1 °C. Once we have the concept of an ideal gas of which Charles' Law is true, we have an ideal scale of temperature. And with the development of the kinetic theory of gases, temperature is not merely a parameter of an ideal gas but is a statistical measure of the mean kinetic energy of the actual molecules of actual gases. By then, it has become fully integrated with the rest of physics. We need have no qualms in forming the concept of a scale of temperature. We can in principle check whether the coefficient of expansion of mercury or alcohol is sufficiently constant for mercury or alcohol thermometers to be satisfactory for measuring temperature with. And we have a clear concept of a Kelvin, and can say why we think the rise from 273 K to 274 K is the same as the rise from 372 K to 373 K.

Temperature is illuminating because it is relatively remote from the central concepts of physics. We need to go through the comparable assumptions implicit in our methods of measuring other magnitudes. In order to establish a genuine *scale*, rather than a mere register of degrees, we need to have some means of establishing equality of differences. We need to have a rule telling us that the increase in weight resulting from adding one gram to three grams already in the balance is the same as that resulting from adding the same one gram weight to ninety-nine grams already in the pan. It is necessary not only to be able to compare different things, and say, e.g., that this thing is heavier than, or hotter than, that thing, but to be able to compare different differences, and say that this is heavier than, or hotter than, that *by the same amount as* a third thing is heavier than, or hotter than, a fourth. More generally, we need to be able to tell when the difference between the magnitude of a and the magnitude of b is the same as the difference between the magnitude of c and the magnitude of d; in symbols:

$$[m(a), m(b)] = [m(c), m(d)]$$

Having found an equivalence relation $\approx$, and formed the quotient set $X/_{\approx}$ and obtained a strict ordering $<$ on it, we need not only to establish a relation, **R**, on that, but another equivalence relation between the **R**-pairs. It seems very complicated. But in practice it often works out simply. Instead of writing the difference between the magnitude of a and the magnitude of b by the general difference relation $[m(a), m(b)]$, we write it with a minus sign; and our requirement of there being some equality between different differences then becomes

$$m(a) - m(b) = m(c) - m(d).$$

This equation follows from

$$m(a) + m(d) = m(b) + m(c)$$

provided we have some addition rule for magnitudes. And this we often have. Thus, in the paradigm case of weight, we assume that weight is additive, and that the weight of two bodies is the sum of the weights of each of them separately. This assumption seems so natural that it is commonly taken for granted. But we should be wary. Consider electrical resistance. Is it unquestionable that the total resistance of two resistances is the sum of their resistances separately? We are tempted to say Yes, until we remember that resistances can be connected together not only in series but in parallel. Indeed, if we had called them conductors, and asked the effect of adding their conductances, the very change of words would probably have suggested that they were to be connected in parallel. Our two addition rules for resistances depend on two assumptions underlying our understanding of Ohm's laws, viz. that where resistances are connected in series the current flowing through each is the same, and where they are connected in parallel, the voltage across each of them is the same. There is no reason to doubt these assumptions—indeed much reason to believe them correct: but they are not self-evident, and ought therefore to be made explicit.

Although it is sufficient for establishing a scale that we should have an addition rule, it is not necessary. Once again temperature is an example. There is no physical operation by means of which we can "add" degrees. Although if a body is at 99 °C I can raise it by 1 °C to be at 100 °C, and can reckon that the difference between 99 °C and 100 °C is the same as the difference between 0 °C and 1 °C, I cannot add the temperature of a body at 0 °C to the temperature of one at 100 °C, or that of a body at 1 °C to that of one at 99 °C. My belief in there being a scale of temperature depends not on a procedure for adding temperatures, but on a theory which enables differences of temperature to be compared, which in turn rests on there being other magnitudes which can be added. We cannot add Kelvins, but we can add calories, or, more fundamentally, kinetic energy. This justifies our regarding temperature as a scale, and also shows that Kelvins could not be easily added, because they are a measure of mean kinetic energy, and mean kinetic energy is a ratio— total kinetic energy divided by number of molecules—and ratios do not add easily. It is the same with density and with pressure. Each is a scale, but there is no natural analogue to addition because if we are to compound two temperatures, two densities, two pressures, we need to identify them by two material objects that have them, and if we then consider the two as one, we alter the denominators of the ratios involved. Only with considerable care can I add a specific amount to the temperature, density or pressure of a specimen, just as only with great difficulty can I add to the atomic number—the number of protons per atom—of an atom.

Campbell places great emphasis on the distinction between measurements where there is a process of compounding by which the measured magnitudes are added—such as mass, length, force—and those where there is not.[5] But the distinction is not as important as

[5] N. R. Campbell, *Foundations of Science*, pp. 275–83.

he makes out. What is important is that where the magnitude to be measured cannot be directly added, it can be expressed in terms of others that can. There must therefore be some subset of magnitudes in terms of which all others can be expressed, and which themselves can be added directly. But the fact that a magnitude can be expressed in terms of others does not mean that it cannot itself be added directly according to some procedure of composition. Current can be expressed in terms of the ratio of the potential difference and the resistance, but can be added nonetheless. In Newtonian mechanics mass, momentum and velocity are interdefinable, but each can be compounded according to its own addition rule.

Campbell also conflates the distinction between fundamental and derived magnitudes with the distinction between a substance and its properties. Mass, weight, and size are, on his view, characteristic of a particular specimen, and are subject to a simple addition rule, whereas density is characteristic of the generic type of substance, and is not subject to a simple addition rule. But temperature and pressure, being ratios too, are not subject to the addition rule, but are not characteristic of types of substance either, and kinetic energy, electrostatic charge, electric current, voltage and resistance, which can be added directly, do not seem all to measure the substance rather than the properties of things.

If we can define addition we can define scalar multiplication. If we can say that the difference between the weights of a and of b is the same as the difference between the weights of c and of d, we can say, too, whether it is twice as much as that between the weights of c and of e. For we could find two other things, f and g, such that $m(f) = m(g) = m(c) - m(e)$. We should put e on one side of the scales and c on the other and then by successive approximations find an object—which would in practice be a collection of weights—that when placed in the same pan as e would make it balance against c on the other side. Removing c and e from the pans we could similarly find another object g that would balance against f. And then placing these two on the same side together with b, we could see whether a balanced against them or not; and if it did, then

$$m(a) = m(b) + m(f) + m(g),$$

so

$$m(a) - m(b) = 2m(f) = 2[m(c) - m(e)].$$

It is all very obvious, but it is worth checking through to make sure that given a rule of equality and a rule of addition we have a rule of scalar multiplication. It is clear, then, that we have a rule of division too, so that we can, within the limits of accuracy of our methods, assign a positive rational number to the ratio of the weights of any two bodies; or, taking any one of them as a unit, assign a positive rational number as measuring the weight, in terms of that unit, of any other.

We have taken weight as a paradigm of a measurable quantity, but it is not the most important one in physics. Physicists are more concerned with mass (which they distinguish from weight in theory, although in practice they measure it by measuring weight in a constant gravitational field), and with length, angle and duration. These are measured in ways which are familiar but in need of scrutiny. As with weight, the main task is to establish equivalence relations which pick out equivalence classes all of whose members have the same length, the same angle, or the same duration. This accomplished, we establish an order on the basis, as we have seen, of the appropriate part-whole relationship, on the relatively rare occasions when it holds, and an addition rule under similarly stringent conditions. If three points, *A*, *B* and *C*, lie on a straight line, then the sum of the distances *AB* and *BC* is the distance *AC*. If the lines *OP*, *OQ* and *OR* are coplanar, then the sum of the angles *POQ* and *QOR* is the angle *POR*. If one process ends when another begins, the duration of the whole process is the sum of the durations of the separate ones. These rules are simple. But they are conditional, and often the condition is not fulfilled. Most processes are not such that when one ends the next immediately begins. If I am adding up my hours of work, I am adding intervals that do not all adjoin one another. Equally when I am measuring up my new house for curtains, I cannot count on all the windows being contiguous, and together forming one vast window area. Hence the addition rule will not apply as it stands, and, in order to be able to apply it, we have first to map equal durations or equal distances onto some special case where the addition rule will apply. Because we believe in clocks, and believe that one hour told by a clock is equal to another hour, I can map each of my hours of work onto successive unbroken intervals of time, and then add them all up. Because we believe in rulers, and believe a yard measured by my tape measure is equal to a yard of curtain material measured out in the shop, I can buy a continous length of curtain material, and reckon that the length of the whole is the sum of the lengths of adjoining sections. We can operate with a relatively stringent condition for applying the addition rule because we have relatively relaxed conditions for establishing equality of duration or of distance, and therefore can transform cases which do not fall under the conditions for the application of the required addition rule into equivalent cases which do.

The way in which we establish an equivalence relation for length, for angle, and for duration, depends partly on empirical factors, partly on a *a priori* assumptions much as did the way we established an equivalence relation for weight. We lay off lengths and angles with rigid rulers and protractors, and are not embarrassed by finding, as conceivably we might, two rigid bodies being equal to a third body and not to each other. Just as we take care not to weigh iron in a magnetic field and to seal up volatile substances such as camphor and ether, so we do not use rubber for making rulers or dough for making protractors. We also are assuming that the length and angle of a rigid body are constant over time and are not altered by a change of position or orientation, just as we assume that the weight of bodies does not alter significantly with the elapse of time, or when moved short distances, or when placed on scales. These are the assumptions of irrelevance we have already needed to make in order to operate inductive arguments and apply the concept of cause, but they are none the less strong assumptions to make, and impose a stringent condition on the structure of space.[6]

The assumptions implicit in laying off lengths and angles with rigid bodies become clearer if we contrast them with duration. We cannot move durations in time. We cannot lay off one temporal interval against another. We have to assume that certain intervals are equal—isochronous. We assume, in fact, that the intervals taken by different instances of the same type of process are the same: complete rotations of the earth on its axis, complete orbits round the sun, oscillations of a pendulum of a given length, vibrations of crystals of quartz, or of photons emitted by caesium atoms. The assumption is one which could be questioned. Often it is suggested that we meet such questions by stipulation, and stipulate that the intervals taken by different instances of the same periodic process are to be *deemed* equivalent as regards their temporal duration. It is sometimes made out that this is as much a matter of convention as

[6] See further below, Ch. VIII, pp. 110–14.

the choice of the second as a unit. But there is an important difference between the choice of a unit and the stipulation of isochronous intervals. Of course, we can lay down that any set of clearly marked out temporal intervals be deemed to be isochronous as we can lay down that any set of clearly marked out spatial intervals be isometric. A man too much brought up on Mercator's projection might believe that metre rulers orientated East and West suffered an expansion when moved toward the North and South Poles, and therefore marked out distances which were really more than a metre in length. Many such idiosyncratic grids of isometric intervals could be established. But each would have to be established in some systematic way: we should need a *rule* to tell us what equal lengths were at different latitudes—e.g. how far a soldier could march with a heavy pack between sunrise and sunset—and we should face questions why we supposed that the rule chosen was a good rule.

Many philosophers of science are conventionalists, and answer the question by saying that it is simply a convention what rule we adopt. But this is a mistake. Conventions arise when there is nothing to choose between one rule and another—e.g. driving on the left and driving on the right—but it is important that we all do the same thing. There is no reason why $3 \times 4 + 6/2$ should be read as $(3 \times 4) + (6/2)$ rather than $[3 \times (4+6)]/2$. It is simply that if we can agree on either rule, we can save cumbersome bracketing, and we have, as it happens, agreed on the former rule. Thus it *is* a convention, granted our usual assumptions about space and time, that we take the metre, the degree and the second as units. We could equally well have chosen the furlong, the radian and the fortnight. We need to have some standard units, and we have found it convenient to settle on these. It is very different with the choice of a rule for picking out isometric or isochronous intervals. A deviant rule would not only be inconvenient: it would be wrong. We have an ideal of accuracy which goes far beyond that of practicality, and witnesses to our being guided by theoretical considerations of truth, and not only pragmatic ones of convenience.

The importance of theoretical considerations is most evident in our measurement of time, because the assumption of isochronous intervals is obvious, and obviously open to question. Alfred's candles, the waterclock of antiquity, and the interval between sunrise and sunset, are poor ways of measuring time because the process of burning varies with draught, the process of running out of a tank

depends on the height of water remaining in the tank, the interval
between sunrise and sunset varies with the season of the year. The
pendulum was first noticed by Galileo as correlating empirically with
his heart-beat, but was adopted as a basis for measuring time only
when it was understood that for small angles the movement of a
pendulum would be in simple harmonic motion with period dependent only on the length and gravitational field. In order to eliminate dependence on the condition that the angle be small, Huygens
devised the cycloidal pendulum. In order to eliminate variation of
length with temperature, compensating bars were devised whose
overall length would remain constant over a wide range of temperature. In the same spirit astronomers replaced the interval from
sunrise to sunset by the interval from noon to noon, then the mean
solar day, then the mean sidereal day. Even the mean sidereal day
is not quite accurate. It depends on the conservation of angular
momentum, but the angular momentum of the earth is being slowly
dissipated by the tides. The quartz clock and the caesium clock
depend on quite different principles—those of quantum mechanics—
and are thought to be more accurate because they are not subject to
the theoretical imperfections of the pendulum and the earth, and
because quantum mechanics is thought to be true.

That quantum mechanics be true is not the only assumption
underlying our measurement of time. We are assuming also that
atomic clocks are in step with ideal mechanical or astronomical
ones—that if today there are 9,192,631,770 periods of vibration of
the outermost electron in an atom of caesium-133 in the period of a
second pendulum or in the 31,556,925.9747th part of the tropical
year, then the same ratio would hold good at all other times.[7] This
assumption has been questioned. Milne has suggested that besides
the ordinary t-time there is another τ-time in comparison with which
the periodic processes of t-time are, in fact, speeding up.[8] Equally,
the assumption behind our use of rigid rulers and protractors to
measure out equal distances and equal angles can be questioned.
Although in Newtonian mechanics and the Special Theory of Relativity we assume that space is "flat" and therefore of constant curvature, we do not assume that it has constant curvature in the General
Theory of Relativity, and therefore are not entitled to believe that
the use of rigid rulers and protractors will be absolutely accurate—

[7] For a fuller account see G. J. Whitrow, *What is Time?*, London, 1972, Ch. 4.
[8] E. A. Milne, *Kinematic Relativity*, Oxford, 1948.

although, it should be added, the variation of curvature in any region we are interested in is so small as to lie within the limits of experimental error.

The assumptions lying behind our equivalence relations for equal distance, equal angle, or equal duration, are thus in need of careful examination and justification. Once such equivalence relations are established, the rest is, as we have seen, plain sailing. By comparing lengths with equivalent lengths on a ruler we can say which is greater than which: by taking adjoining intervals on the same ruler, each being equivalent to a length we are interested in, we can add them together. Under these stringent conditions the addition rule becomes very easy. Similarly with angle and duration.

A more difficult case occurs, however, with velocity in the Special Theory of Relativity. We might take it as self-evident that if one body had a velocity u with respect to a given frame of reference which itself had a velocity v in the same direction with respect to another frame of reference, then the velocity of the body with respect to that other frame of reference would be.

$$u + v.$$

But this turns out not to be the case. The rule that the Special Theory of Relativity requires is

$$\frac{u + v}{1 + \frac{uv}{c^2}}$$

where c is the velocity of light. This will give a lower total than the simple addition rule. In particular it will give the highly counter-intuitive result that if either or both of u and v are themselves equal to c, the resulting velocity will not be increased at all. Thus, if $v = c$, we have

$$\frac{u + c}{1 + \frac{uc}{c^2}} = \frac{u + c}{1 + \frac{u}{c}} = \frac{c(u + c)}{u + c} = c.$$

Hence in this special case, even the ordinal properties presupposed in a theory of measurement are lacking. This might seem a serious blemish, but will look less bad if we reflect that even with the ordinary addition rule, $\pm \infty$ behaves in a similar way. If we regard the velocity of light as being an infinite velocity, then it would naturally have this property. But can we? Let us define a new physical quantity, call it "rapidity", by the equation

$\varphi = \tanh^{-1}(u/c)$, and correspondingly
$u = c \tanh \varphi$.

We then find that the addition rule does hold for rapidities, and that the complicated rule for velocities is just what comes from our transforming the simple rule of addition for rapidities into velocities; that is,

$$\tanh(\varphi + \psi) = \frac{\tanh \varphi + \tanh \psi}{1 + \tanh \varphi \tanh \psi}.$$

We can give a sort of picture of rapidity. If we were using the ordinary tangent, not the hyperbolic one, then $\tan\phi$ gives the gradient in terms of the angle the line makes with the X-axis. We can regard the hyperbolic tangent as being a comparable function of a pseudo-angle. So we could make out that the really important physical feature was the pseudo-angle, measured by rapidity, not the familiar quotient of distance by time. Should we do this? Can we always do this? The answer to the first question in this case is a tentative No, to the second a surprising Yes. It turns out that under a surprisingly wide range of conditions we can "regraduate" any magnitude so as to reduce its composition rule to a simple addition rule. We require that the magnitude be one-dimensional, and that the composition rule be differentiable, monotonic and subject to the associative law. This excludes vectors, but little else in physics. Hence we generally can choose our scale so that the addition rule will apply. From this it might seem to follow that the selection of physical magnitudes was very largely a matter of *fiat*: energy just is that scalar magnitude of systems which can be simply added. But although *a priori* considerations are important here, as elsewhere in physics, they are not all-important. Although we *can* define a physical quantity as that which some physical operation adds, it may not lead to further understanding. In the case of the Special Theory of Relativity, the use of pseudo-angles is illuminating indeed, but not fundamental. We therefore do not replace velocity by rapidity, but rather retain

---

[9] I am grateful to M. J. Lockwood, of Exeter College, Oxford, for introducing me to the term. For further discussion of angle and pseudo-angle as a measure of velocity, see E. F. Taylor and J. A. Wheeler, *Spacetime Physics*, Freeman, San Francisco, 1963, §9, pp. 47–58.

velocity as our fundamental concept, accepting a somewhat complicated non-additive rule for the composition of velocities. If we are to measure scalar magnitudes at all we must have some composition rule, and tend to measure them in such a way as to have our composition rule a simple addition rule. Thus we measure both resistance and conductivity, which can be added by putting in series and in parallel respectively. But this is not the only consideration. It is no use measuring a magnitude unless the measurement will be useful to us in our daily affairs or illuminating in our attempts to understand the nature of things.

The choice of an addition rule determines a zero and leads to a choice of a unit. Often the choice of zero is dictated by the ordinal or other properties of the magnitude being measured. If there is a temperature lower than which it is impossible to get, then that is the temperature which should be taken as absolute zero. Once we have granted that distance is

(1) a function of the two points between which it is being measured,

(2) a symmetrical function—i.e. that the distance between A and B is the same as that between B and A,

then the distance between A and A is clearly a limiting case; and if our ordering relation is based on inclusion, the distance between A and A is clearly nought, corresponding to a null interval. Thus, often we have a natural zero.

The choice of a unit is seldom dictated by the nature of the concepts, and not all that often by the facts. The best example where it is thus dictated is in the unit of angle. If something is rotated through 360°, it comes back to pointing in the same direction. Hence one revolution, which the trigonometric functions lead us to construe as 2π radians, is a natural unit of angle. Granted a certain Riemannian geometry, there would be a natural unit of length, as there is on the surface of a sphere, where great circles are the longest possible "straight" lines. Thus topological considerations can yield metrical results; and so, conversely, if we make the metrical assumption of there being no naturally given unit, we are committing ourselves to only a limited range of possible topologies.

In the Special Theory of Relativity we take the speed of light as a natural unit. More contingently, Planck's constant offers itself as a natural unit of action, in so far as we might wish to measure that aspect of a physical system. Equally contingently, the charge of the electron is a naturally given unit, and perhaps also its mass. Granted some naturally given units, we could define others. If there were a maximum length, we could define a "great year" as the time a photon would take to go right round the universe. Or we could define time in terms of action, speed and mass, and so a unit of time in terms of those naturally given units. But such units at present do not seem to have any real significance. This may be due to our lack of understanding. There may be a natural unit of duration which explains why there is a fundamental rhythm in the universe and all natural processes keep in time. But we are far from understanding this, and at present we can only regard duration and distance as having no natural units; and although the mass of the electron—or the neutron—is naturally given, we are not yet sure of its fundamental significance.

TABLE 1

	Weight	Duration	Length	Angle	Velocity	Resistance
Equivalence Relation	balance	taken by similar processes	laying off against rigid body (ruler)	super-imposing by rigid body (protractor)	keeping up with	being of same substance and same dimension
Strict Ordering	includes	includes	includes	includes	overtakes	being a serial part of (as in a potentiometer)
Addition Rule	lump together	adjoin	adjoin in same direction	adjoin in same plane	derivative magnitude—complicated rule	arrange in series
Natural Zero	Yes	Yes	Yes	Yes	Yes in Newtonian Mechanics No in Special Theory of Relativity	Yes
Natural Unit	No	No, at least not yet	No	Yes	No in Newtonian Mechanics. Yes in Special Theory of Relativity	No

Exercise. Construct a similar table for other magnitudes—mass, acceleration, force, energy, temperature, conductivity, voltage, etc.

Exercise. How far can we fit into this table the degenerate cases where we assign numbers to positions or dates as co-ordinates?

We can summarise the way we apply our various metrical assumptions to measure different magnitudes in a table (see Table 1). It is worth considering how the table would look for other magnitudes—temperature, value, force, voltage, etc. It is also worth considering the case of the assignment of co-ordinates to dates and places. This, like a proper measurement, needs to be interpersonal, and to that extent nonarbitrary, but essentially has no addition rule, and no unit. There is an equivalence class and a strict ordering for non-relativistic dates. In non-relativistic theories we have an equivalence relation of simultaneity which picks out all those events which happen at the same date, and these dates can be strictly ordered by *before* and *after*. There may be a natural zero, although it is not generally agreed to be a conceptual necessity. With space the position is more complicated. In common-sense—Aristotelian— conceptions of space we can assign different events to the same place because we have a workable criterion of absolute rest. In Galilean and Newtonian physics, an inertial frame need not be at rest absolutely, and so we have some difficulty as physicists in knowing how to apply the equivalence relation *in the same place as*. We need to distinguish the two possibilities, which we might name Leibnizian, in which we cannot apply *in the same place as* to events at different dates, and Clarkean, in which we can. In both cases further assumptions about parallels are normally made, which, if allowed, enable us to establish a strict ordering in each direction. In the Middle Ages it was sometimes maintained that there was a natural zero, namely Jerusalem; but our commitment to the homogeneity of space leads us to discountenance this, and to say either that there is no natural origin in space at all, or that if there were it would not be of any physical significance.

Further Reading

A   David Bostock, *Logic and Arithmetic*, vol. 2, Oxford, 1979, esp. Ch. 2, §4, pp. 112–30. (Difficult but rewarding.)

C   Brian Ellis, *Basic Concepts of Measurement*, Cambridge, 1968, Chs. IV, V and VIII.

Preliminary Reading for Chapter VII

A   P. C. W. Davies, *The Forces of Nature*, Cambridge, 1979, Ch. 5, §5.2.
B   David Singmaster, *Notes on Rubik's Magic Cube*, London, 1979, pp. 4–10.

VII

Constancy, Invariance, and Symmetry

H UME elucidated causality in terms of constant conjunction, and we have seen that it is an essential feature, although not the only one. But it needs further elucidation. 'Whenever A then Z' sounds simple enough, but in addition to the complexities lurking in A, there are more fundamental ones in the 'whenever'. What we observe are instances, *this* A, *that* A and *the other* A. As instances they are all different—in Chapter IV we symbolized them as A(1), A(2), A(3), *etc.*,—but we are concerned that they are all instances *of the same type*. We cannot perceive types—universals—directly: even Plato, who laid the greatest stress on universals, thought they could be seen only by the eye of the mind. Our actual bodily eyes can only see instances—this A-thing, that A-thing, and the other A-thing, or, symbolically, A(**1**), A(**2**), A(**3**). But what we are interested in is their all being instances of A—symbolically **A**(1), **A**(2), **A**(3). Although observations and experimental results are crucial for the natural sciences, scientists have a very different attitude to them from historians. Historians are interested in what happened in all its concrete individuality. A historical account of the Michelson–Morley experiment or the discovery of DNA would specify the exact date and place, who actually was present, who made what suggestions, even perhaps the sort of biscuits provided that day for lab tea. The scientist cares for none of those things. He affects a deeply impersonal style, and ignores irrelevant circumstance. He is concerned only to report results relevant to the confirmation or refutation of possible scientific hypotheses, and is not concerned with the actual instance in all its particularity, but only with its being a typical instance, and proof therefore that some specific combination of types is, or is not, instantiated. Instances, that is, are important not in themselves, but only in order to decide whether or not types are instantiated. Nevertheless, we are obliged to have them, and they introduce an element of ineliminable difference into our data, which we need to discount, but cannot entirely ignore. No two instances are the same in absolutely all respects. At the very least they differ in respect of spatial or temporal positions, else they would be one and the same instance. Ideally they may differ in no other respect, and, as we have seen,[1] the fundamental doctrine of chemistry is that there are chemical substances which are qualitatively identical in every way except that of occupying different spatio-temporal positions. In practice, instances often are different in some further respects, and even when the only differences are spatial or temporal, we may not be entitled to assume at the outset that all such differences are irrelevant. Although as seekers after scientific truth our concern is with underlying samenesses, yet as practitioners of scientific method we have to base our enquiry on given instances which are always in some respect different. Absolute sameness is unattainable: it is always some sort of similarity, a sameness in some respects but not all, that we seek.

The concepts of being *the same as, similar, like, unlike, different from, dissimilar*, and of *resembling* are relations, but are three-term relations. If you ask me whether I am the same as you or different from you, I cannot answer until you say in respect of what. I am the same as you in respect of being a man, being interested in philosophy, being able to understand English. I may be the same as you in nationality, education and hobbies. I am probably different from you in respect of height, complexion and name. We both share a common humanity, but we do not both have the same name. You need to specify whether you mean 'same in respect of being human' or 'same in respect of nomenclature'. So, too, in physics, two bodies may have the same mass, but different kinetic energies, the same speed, but in different directions, emit light of the same frequency but of different intensities. We cannot simply observe a number of events which resemble one another in being constantly conjoined with other events, themselves all resembling one another too. Resemblance is not, as Hume assumes, an immediately given relation

---

[1] Ch. III, p. 39.

but depends on schemes of classification, which are themselves complex and subject to amendment. We have to learn how to classify things as similar or dissimilar. It is a large part of learning a language to be able to discriminate things and experiences according to the different linguistic descriptions available. Even with colours, where Hume is on strongest ground, different cultures distinguish different hues and shades. The ancient Greeks had a very different colour vocabulary from ours, and had no precise equivalents to our blue or green or yellow, and the Eskimos are said to have many different words for different sorts of white. Taxonomy is an important part of biology. The biologist has to learn how to classify before he can go on to do anything else. It is only by careful training that the ethologist can learn to recognise and describe significant forms of behaviour. This is not to say that our recognition of resemblance is altogether a result of learning or linguistic competence; animals can recognise certain sorts of similarity. But it is to say that similarity is not a simple dyadic relation nor entirely given in experience. Often we are best at discerning rather complicated patterns—we can tell at a glance whether a man is angry or not, but only a skilled artist can describe the exact configuration of his face or stance—and our descriptions depend in part on our interests and theoretical presuppositions. The same holds good in the physical sciences. Our criteria of resemblance depend in part on our theories. Aristotle took it as obvious that the circle and sphere were significant shapes, whereas we often will consider an exponential or sinusoidal correlation more profoundly important.

We can handle instances which are characteristically different, while seeking the underlying samenesses they manifest, if we make use of the theory of groups. Groups have been made familiar to many by means of Rubik's Magic Cube. The six faces of the Magic Cube are each divided into nine coloured squares, which are so interlocked that they can move independently, thus allowing any one face to be rotated with respect to the rest of the cube. If a face is turned through 90°, 180°, 270° or 360° in either direction, the overall shape will once again be cubical, although the arrangement of the square faces may be different. Any sequence of such turns of any face will re-arrange the components of the cube in some way or other. We call such a rearrangement a TRANSFORMATION. It is clear that one transformation followed by another transformation is itself a transformation, and that if we have a sequence of three transformations in a given order, it does not matter whether we regard it as the result of combining the combination of the first two, taken together, with the third, or as the result of combining the first with the combination of the second and third taken together.Thus if I turn the right face 90° clockwise, and then the upper face 180°, that will be a transformation of the cube too; in Singmaster's notation[2] it will be the transformation RU^2; and if I then turn the left face 90° clockwise, the result will be $RU^2 L$, and we do not have to put brackets in, and distinguish $(RU^2)L$ from $R(U^2 L)$, in the way in which we *do* have to distinguish $(3^3)^3$ from $3^{(3^3)}$. It is also clear that to every transformation there is an inverse transformation which undoes the effect of the first transformation and takes the cube back to exactly what it was before. Thus if I turn the right face 90° clockwise and then turn the right face 90° anticlockwise, the result is to leave the cube in the identical state that it was before. We indicate an inverse by writing a -1 superscript. Thus if R is a 90° clockwise turn of the right face, R^{-1} is a 90° anticlockwise turn of the right face. Somewhat contrary to ordinary English usage, we regard leaving the cube in the identical state that it was as being itself a sort of transformation, namely the Identity transformation, which we indicate by I. Thus we write $RR^{-1} = I$. It is obviously also true that $R^{-1} R = I$. There are many other sequences of transformations that result in the Identity transformation, for example $R^2 R^2 = I$. We should note that although with Rubik's Magic Cube it is obvious that each transformation has an inverse, such a rule does not hold for other operations. Thus I can disjoin two propositions, p and q, to form a new proposition $p \vee q$, but I cannot then de-disjoin q to get back to p: there is no $(\vee q)^{-1}$ such that $(p \vee q)(\vee q)^{-1} = p$, and the fact that there is not made our account of the logic of cause and effect in Chapter IV much more cumbersome than the corresponding account of functional dependency[3] where the effects of additional causal factors can be subtracted from the total effect to yield the effect due to the one we are interested in. We have to lay it down, therefore, as part of the definition of a group, that there is an Identity transformation and that every transformation has an inverse. But although in the special case of a transformation and its inverse we have $RR^{-1} = R^{-1} R$, it is not in general true that the order of transformations is immaterial. It is easily seen on the Magic Cube that $RU \neq UR$. That is to say, transformations are not, in most

---

[2] David Singmaster, *Notes on Rubik's Magic Cube*, London, 1979, p. 4.
[3] See below, Ch. X, pp. 161 ff.

groups, commutative. A group in which transformations are all commutative is called an Abelian group. If I confine myself to one face of the Magic Cube, then I have an Abelian group; thus $RR^2 = R^2 R$.

The aim with the Magic Cube is to be able to transform it so that all the squares on each face are the same colour. It is not too difficult to get all the squares on one face the same colour, but then almost any transformation which gets another face correct puts the first face wrong; thus if I have got the left face correct, the only simple transformations which leave it correct are R, R^{-1} and R^2, and these are not enough to move all the other faces to their correct position: if I am to move them, I must disturb the left face temporarily, and then undo the disturbance I have caused. Thus I can alter the top and right faces, leaving the left face ultimately the same, by the sequence $U^2 R UR^2 U^2$. There is a great premium on finding transformations which, when they are complete, will leave the one face that was already correct still correct; such transformations form a subgroup, that is, a group contained in the original group, and the unaltered face is an INVARIANT of the subgroup.

There are many other examples of groups in ordinary life—bell-ringing, arranging guests around a dinner-table—and, especially in physics, we need to consider infinite as well as finite groups. Formally we say that a group is constituted by a set of "elements", A, B, C, etc., and a binary connective, $_o$, subject to the four rules:

1. The result of combining any two elements of a group is itself an element of that group.
2. Elements combine according to the associative law; that is to say
    $A \,_o (B \,_o C) = (A \,_o B) \,_o C$.
3. There is an identity element, I, which for any element of the group, A, has the property
    $A \,_o I = A = I \,_o A$.
4. Every operator, A, has an inverse, A^{-1}, such that
    $A \,_o A^{-1} = I = A^{-1} \,_o A$.

The formal characterization is entirely abstract, but in its applications, group theory is concerned with transformations. These may be regarded "actively" as operators, which actually alter something—as when we alter the arrangement of the Magic Cube—or "passively" as redescriptions—as if we looked at the Magic Cube

from the right instead of the front. In either case they are transformations *of* some entities, e_1, e_2, e_3 . . . e_j. . . etc.—which might be
squares on the cube, soundings of bells, dinner guests, points in
space, moving particles, rigid rods, *etc.* An invariant feature is some
function $f(e_k, e_l, e_m, \ldots)$ which is such that for any member of the
group, g_i

$$f(e_k, e_l, e_m, \ldots) = f(g_i(e_k), g_i(e_l), g_i(e_m) \ldots).$$

In this way group theory enables us to express identity in difference. Each transformation makes a difference. Each transformation is different, and makes a different difference. But we can
recognise an underlying sameness none the less, and express it in
terms of invariance under the group of transformations, and can
express even more subtle interplays of sameness and difference by
means of different subgroups within one overall group. Consider
the group of displacements, reorientations,[4] reflections and magnifications. They may be regarded actively as operators—we move
something along, turn it round, turn it over, or sometimes make it
larger or smaller—and then clearly they alter positions, directions
and sometimes sizes, but do not alter shape. They may alternatively
be regarded passively, as being not alterations in the things themselves, but only in the way we regard it or describe it. But then again,
I do not alter the shape of a thing, if I look at it in a mirror or
through a magnifying glass or a telescope the wrong way round.
Either way, shape is invariant under this group. Size is not invariant
under this group, but is invariant under the group generated by
displacements, reorientations, and reflections; and congruence is invariant under the group generated by displacements and reorientations. These latter two groups, known as the Euclidean group
and the proper Euclidean group respectively, leave invariant the
properties studied in Euclidean geometry. Other geometries can be
characterized as being invariant under other groups.[5] The most

---

[4] Many writers use the terms 'translation' instead of 'displacement' and 'rotation'
instead of 'reorientation'. The word 'rotation' suffers from a confusing ambiguity,
in that it can refer to something's having rotat*ed* through an angle, or to something's
rotat*ing* with an angular velocity. I therefore sacrifice idiom and ease of expression
to clarity, and use only the awkward terms 'reorientation' and 'angular velocity'
throughout.

[5] The group-theoretical approach to geometry is due to Felix Klein, who proposed
it in 1872 in his *Erlangen Programme*. Although very different from the traditional
axiomatic approach of Euclid, it gives a much better grasp of the application of
geometry to the natural world.

general of these is the group of one–one continuous transformations, and we can define a *topological* property as one that is invariant under the group of one–one continuous transformations. The larger the group, the more general the property; and conversely, the more specific the property, the smaller the group of operations that leave it unaltered. Physicists take great care in their experiments not to disturb the system in certain respects—to leave its total mass, energy, heat, or electric charge, unaltered. They are, in effect, confining themselves to a limited subgroup of operations, over which mass, energy, heat or electric charge are invariant.

The proper Euclidean group is important because, as agents, we move things around, putting them in different places and facing different ways, and as spectators, we move around and see things from different positions and perspectives. In our description of the world we are thus impelled to pick out features which are invariant over the proper Euclidean group. Shapes and sizes are markedly different from colours. Locke classified them as primary qualities, and held that they really existed in things rather than being merely produced by them in us. But this is partly due to the exigencies of perception and communication. If we did not pick out properties that were invariant under displacement and reorientation, we should be unable to recognise as the same what we see from nearby and from afar off, from one side and from the other, and should be unable to talk to one another about it. Consider two people looking at a penny lying on a table. It will have an elliptical appearance, with the ellipse having a different eccentricity for each of them (unless it happens to be symmetrically placed between them). If they were to describe the ellipse they each saw, they each would describe something different. The only way two people who are necessarily not in the same place at the same time can talk about things which are not in general symmetrically placed between them is to talk about not the elliptical appearances of the penny, whose eccentricity is different for speaker and for hearer, but the round reality which is equally circular for both. Even one person on his own would be impelled to notice and remember not the apparent shape which varies from position to position, but the feature which is invariant under changes of position and perspective, and to regard this as the real objective shape.

So great is the pressure that the necessities of communication and of recognition exert on our minds that not only do we have to talk about real shapes rather than apparent shapes, but we see them. Psychologists have discovered "The Phenomenal Regression to the Real Object";[6] even when we try to concentrate on apparent angles or apparent shapes, our eyes see them more as they really are than as they should, according to the laws of perspective, appear to be. Although the penny looks elliptical, if we are asked to choose from a selection of ellipses of varying degrees of eccentricity the one that most closely matches the apparent shape of the penny, we choose an ellipse which is less elliptical and more round than the retinal image of the penny is. Even when we try not to, we re-interpret the visual stimuli as seeming to be something more invariant than they actually are. We cannot be phenomenalists even when we try, but are naive realists at heart, and cannot help attending to those features that are invariable from place to place and person to person rather than those that are variable and subjective.

The reader can confirm the psychologists' findings by a simple experiment. Let him look up at the four corners of the ceiling of his room and judge what the apparent angle at each corner is; that is, at what angle the two lines where the walls meet the ceiling appear to him to intersect each other. If the reader imagines himself sketching each corner in turn, he will soon convince himself that all the angles are more than right angles, some considerably so. And yet the ceiling appears to be a quadrilateral. From which it would seem that the geometry of appearances is non-Euclidean, with the angles of a quadrilateral adding up to more than 360°. And so it is; but it does not worry us, because we never think of it, hardly ever notice it. It is quite difficult to elicit from most people the answer that the angle appears to be more than a right angle. Asked simply what the angle seems to be, they will say, immediately and simply, "A right angle, of course". The geometry of appearances is almost untalkable about. We cannot refer to apparent angles and apparent shapes, except by artifice and subsidiarily, just because when talking to each other we are necessarily occupying different positions from which things characteristically look different. We have to talk about, and are under pressure even when by ourselves to notice, not the apparent but the real shape.

    [6] R. H. Thouless, *British Journal of Psychology*, 21 and 22, 1931. J. J. Gibson, *The Perception of the Visual World*, Cambridge, Mass., 1950, pp. 169–72. O. L. Zangwill, *Introduction to Modern Psychology*, London, 1950, pp. 30–4. R. S. Gregory, *Eye and Brain*, London, 1966, pp. 152–3.

So shape must be a primary quality. It is different with colour. The colour of objects—apart from a few such as those made of shot silk—does not vary with the point from which they are viewed. The only external circumstance which affects colour is illumination, and this varies characteristically (for our ancestors, at least) only slowly with time. In the course of any one conversation, the illumination will be the same for both speaker and hearer throughout. Therefore both can talk about apparent colours. It was not necessary (until the invention of artificial lighting) to make much distinction between colours as they appeared to be and colours as they really were. Colours could afford to be secondary qualities, in a way in which shapes could not.

The exigencies of communication also require temporal invariance. We cannot both speak at once. If we are to talk to each other, we must talk at different times, and if we are both to talk about the same thing, we must be able to use words at different times to refer to the same thing, and there must be other, descriptive words by means of which we can at different times either say or deny the same things about it. If I am to be able to talk with you about the church tower across the fields, the words 'church tower' must in my mouth at one time refer to the same object as it does in your mouth at another: and if I am to be able to say of it that it has pointed windows, you must be able to mean the same by 'pointed windows' at the necessarily different time when you either concur, and agree that it does indeed have pointed windows, or dissent, and maintain that its windows are, on the contrary, round. This does not mean that we cannot refer to momentary events, like a flash of lightning, or describe things as rapidly changing. But it does lead us to look for relatively long-lasting continuants, by means of which we can anchor our scheme of reference, and to attach significance to features that are invariant over time.

Invariance is not only a mark of intersubjectivity, but of objectivity too. I attach importance to features that remain the same no matter what I do. In my dreams everything may be subject to my wishes, but it is characteristic of waking life that some features are independent of me and recalcitrant to my will. At the crudest level material objects are solidly there whether I acknowledge them or not, and I shall bump into them if I attempt to walk through them. The simplest causal connexions are four-dimensional analogues: I cannot put my hand into the fire without getting burned, or into water without getting wet. Often, however, the interplay of possibility and impossibility is more subtle. I can choose whether I see the starry heavens or not, but if I choose to look up at them on a clear night I cannot see Orion riding on the back of the Bear or see the Pleiades surrounding Alpha Centauri in the Southern Cross. By walking across the field I can keep the church tower in view, but if I go down the lane it will be hidden by the hedge. By acting appropriately I alter my sense experience or the course of future events, but only in certain ways. There are correlations between what I do and what happens, which can be seen as ineluctable constraints that reality imposes on me. Just as I am peculiarly conscious of myself in my freedom of action, so I am made aware of the existence of something other than myself in the limits to my freedom of action. These limits show themselves not simply in crude impossibilities, but in features that remain the same no matter what I do or what else happens. Invariance, and especially invariance over time, is thus taken as a mark of reality, and we accord such features special ontological status, as being real features of the physical world.[7] When it was discovered that the mass and energy of an isolated physical system would always turn out to be the same at the end of a period as at the beginning, it was an indication of the reality of mass and energy. Likewise the discovery that in an isolated chemical reaction the quantities of each chemical element present remained the same although the quantities of other chemical substances might well be altered indicated that the elements were the fundamental entities of which material objects were all made up.

To be an invariant is, in terms of group theory, to be the same. But absolute sameness is, as we have seen, outside the scope of physics, which is concerned with things which are similar in some respects and dissimilar in others. In ordinary language there is uncertainty whether two things should be described as being the same in certain respects, or as being only similar, and there is a corresponding difficulty, reflected in the usage of many physicists, between the concept of INVARIANCE or an INVARIANT on the one hand and that of COVARIANCE or a COVARIANT on the other. Consider once again Rubik's Magic Cube. Suppose it had started with every side a uniform colour, and had been scrambled by someone who had used only those transformations which rotated opposite faces each through 180°. Although none of the cubelets might be in its correct

---

[7] See Ernest Nagel, *The Structure of Science*, London, 1961, pp. 149–52.

position, the relation of each to the others would have varied in so systematic a way that there would be an evident symmetry, and a skilled cubemaster would see at once that he could solve the cube in a few moves. Although no face was left unaltered—invariant—by the transformation, there was none the less some similarity of structure in the relationship of the different cubelets that had been preserved. Often also in describing the external world we want to talk of things being only similar, and not exactly the same. Instead of concentrating on the real shape and real size of objects, as seen from different points of view, we might consider—as indeed we did consider—their apparent shapes and apparent size, and still discern some similarity between the apparent shapes and sizes of a penny as viewed from different angles and different distances. This is what an artist portrays. By use of the laws of perspective, he depicts how things would look if set at a certain angle and distance from the artist's point of view, and although he depicts differently windows which are of the same shape and size but differently situated, they differ in a systematic way, and, without being the same shape or size, none the less manifest some sort of similarity. The angles of the quadrilaterals on the canvas that depict the windows and the eccentricity of the ellipses that depict the pennies vary with the point of view of the artist in such a way as to correspond with his view of the rectangular windows and the circular coins. It is this similarity of correspondence that the concept of covariance seeks to capture in mathematical form. Whereas an invariant is something which remains the same under a group of transformations, what is covariant under a group of transformations is a law or correlation or expression, expressed by an equation, which is similar inasmuch as it has the same form, however the terms are transformed by any transformation of the group. Thus the distance between two points remains the *same* under any transformation of the Euclidean group: it is an invariant; the *rule* for obtaining the distance between two points is expressed in terms of their co-ordinates—

$$d = \sqrt{((x_1 - x_2)^2 + (y_1 - y_2)^2 + (z_1 - z_2)^2)}$$

—and to this extent depends on our choice of co-ordinates; but if we transform those co-ordinates by any element of the Euclidean group, although the rule will then be expressed in terms of the new co-ordinates—

$$d = \sqrt{((x_1' - x_2')^2 + (y_1' - y_2')^2 + (z_1' - z_2')^2)}$$

—and to that extent will be different, it will be *similar* in form, and so will be covariant. Again, if we change the system in which vectors are expressed, we shall change their expression. They are not *invariant* under the change—their components do not in general have the same numerical value in different systems. But the rule for changing them is the same for all. They all change similarly. They transform *covariantly*.[8] Covariance enables us to discount not only differences of circumstances but differences of perspective. In physics we are concerned not only with cases where variations in the cause produce correlated variations in the effect, but also, and in the General Theory of Relativity far more importantly, with cases where variations in the co-ordinate system or frame of reference carry with them corresponding variations in the description of phenomena and laws. Covariance expresses, under extremely sophisticated mathematical shifts of perspective and circumstance, similarity of form, and thus the underlying uniformity of nature.

Invariances are defined by relatively narrow questions. If we ask what the shape or size of an object is, there is a great deal we are not asking about. Where the questions are relatively wide, we talk not of invariance but of symmetry. A system is symmetrical under a group of operators if any operator of that group leaves the system unaltered in all of a large number of relevant respects. That is to say, a symmetry is where some alteration makes no difference. It is inherent in this definition that we are using two different criteria of difference. With regard to one, there is a difference—there has been an alteration—but with regard to the other, there is none—the alteration makes no difference. Thus symmetries occur only when there are two conceptual structures according to one of which there is a difference, and according to the other of which there is none. The alteration is an operator of a group for which every property, or every significant property, other than the operator itself, is invariant. A starfish is radially symmetrical in that if we rotate it through a certain angle, it presents the same outline, with the same organs similarly situated. A chordate is bilaterally symmetrical in that an interchange of one side and the other, right and left, makes no apparent difference. In these cases the symmetry is not perfect, and close inspection or anatomical investigation will reveal asymmetries

---

[8] The words 'covariance' and 'covariant' are also used in a different, although related, sense in tensor calculus, and are there contrasted with 'contravariance' and 'contravariant'.

which are affected by the alteration—the heart is on the left side, not the right. In inorganic nature the symmetries are more perfect, but still we feel a residual sense of ultimate individuality: if we rotate an ice-crystal through 60°, it may be the same in all discernible respects, but still there *is* a difference in that the atoms of hydrogen and oxygen which were here are now there, and moved from the one place to the other by a spatio-temporally continuous path. Without some such ultimately differentiating principle we feel vulnerable to Leibniz's argument for the Identity of Indiscernibles. In quantum mechanics, however, such an ultimate principle of individuation is not vouchsafed to us, and we have to consider carefully whether an apparent difference makes any real difference at all.

Every group of transformations will leave some features invariant and will transform some relations covariantly; and if the group of transformations is physically important the invariant features and covariant relations will be physically important too. As group theory has become more important in modern physics, the approach to invariants has altered. The horse and the cart have changed places. Originally, as we have seen, it was a discovery that the mass and the energy of an isolated system were the same after an experiment as it had been before, but now it can be seen more as a stipulation. We ask of a system what feature is invariant under any time displacement, and identify that with the mass and the energy, which in Relativity Theory we regard as two aspects of one entity—massergy. The conservation of mass and energy, instead of being a discovery, becomes a consequence of our accepting that physical laws hold in the same way at all times and physical phenomena should be essentially the same everywhere. Of course, there is still some empirical content to the law. Even if we believe *a priori* that mass and energy are conserved, we still have to identify them in their various manifestations, and recognise that the energy of a body may be in the form of its spinning, its radiating light or its being hot. Nevertheless, the emphasis is different. Because we are committed to the homogeneity of time, we are committed also to the conservation of something, and although we have to discover what forms the conserved quantities take, the truth of the conservation law rests on the homogeneity of time rather than being itself a brute empirical fact.

The homogeneity and isotropy of space can also be expressed in terms of the theory of groups, and also give rise to conservation laws. To say that space is homogeneous is to say that physical laws hold in the same way at all places and that physical phenomena should be essentially the same everywhere. A spatial displacement should in itself make no difference so far as the physicist is concerned. But if the group of spatial displacements is to leave physics unaltered, there must be some quantities defined by that group which are conserved. In this case the quantities are not scalars, like mass and energy, but vectors, and are the three independent components of linear momentum. In the same way the isotropy of space, expressed in terms of the group of reorientations, gives rise to the conservation of angular momentum. In classical physics, it must be admitted, the derivation of conservation laws from *a priori* doctrines of homogeneity and isotropy seems rather artifical, and should not be cited as a strong argument in favour of the rationalist approach. In quantum mechanics, however, the fundamental importance of symmetries is much more evident, and the argument can be put forward with confidence.

Further Reading

A *Feynman Lectures*, Vol. 1, §§11.1–11.3, 16.1, Ch. 52.

B H. Weyl, *Symmetry*, Princeton, 1952.

B E. P. Wigner, *Symmetry and Reflections*, Bloomington and London, 1967, Chs. 1–3, and pp. 57–62, 74–5.

B Peter M. Neumann, Gabrielle A. Stoy and E. C. Thompson, *Groups and Geometry*, The Mathematical Institute, Oxford, 1982, vol. II, Ch. 19, "The Group Theory of the Hungarian Magic Cube".

C M. L. G. Redhead, "Symmetry in Intertheory Relations", *Synthese*, 32, 1975, 77–112.

C L. Sklar, *Space, Time and Spacetime*, Berkeley, Calif, 1974 pbk. ed., 1977, Ch. V.

Preliminary Reading for Chapter VIII

A   Isaac Newton, *Principia*, Scholium to Definition VIII, reprinted in
H. G. Alexander, ed., *The Leibniz-Clarke Correspondence*, pp. 152-60.

A   Leibniz, letters II, §1, p. 16; III §§1-8, pp. 25-8; IV §§1-20, pp. 36-9;
V §§1-20, 66-73, 76-7, pp. 56-60, 78-81.

A   Clarke, letters II, §1, p. 20; III §§2-8, pp. 30-3; IV §§1-20, pp. 45-50,
V §§1-25, pp. 97-100.

A   Herman Erlickson, "The Leibniz-Clarke Controversy", *American
Journal of Physics*, XXXV, 1967, pp. 89-98.

A   *Feynman Lectures*, I §§16.1-16.2.

B   E. Nagel, *The Structure of Science*, New York, 1961, pp. 208-11.

VIII

Homogeneity and Isotropy

THE principle of the causal inefficacy of space and time was too
crudely stated in Chapter V. Indeed, it would be inconsistent to
stipulate both that all spatial and temporal conditions were causally
irrelevant and that causal connexions must be spatio-temporally
continuous. We need to distinguish the irrelevance of absolute spatial
or temporal position from the relevance of spatial or temporal position relative to other causal factors. It makes a lot of difference if I
move a compass needle nearer to an electric current, or if I boil an
egg for ten minutes instead of three. We need to put forward two,
superficially contrasting, theses, one of absolute indifferentism—
the absolute spatio-temporal location of a system is a matter of
indifference—the other of differential relevance—the only relevant
causal factors at any spatio-temporal position are those in its immediate neighbourhood, which can be expressed in terms of dif-
ferences—δx, δy, δz, δt—or differentials—dx, dy, dz, dt. The
principle of repeatability has to be reconciled with that of locality.
We therefore consider transformations which all preserve differences
in spatio-temporal position as between different parts of the same
system. These are symmetries. Symmetry groups enable us to talk
in the same logical breath both of the spatial and temporal differences that need to be accounted different—duration of immersion

of egg in boiling water—and the differences which can be discounted
as making no real differences at all—whether the egg was boiled
when there was an R in the month or no. Three symmetries preserve
differences of spatial position: displacement, reorientation and reflection. Together they form the Euclidean group. The first two,
which together form the proper Euclidean group, are continuous
groups—a displacement or reorientation can be as small as we
please—and they therefore have special physical significance, because they represent not only important similarities of the same thing
under different conditions of observation, and hence also similarities
between one thing and another, but also possible *motions* of the
same thing. A thing, if we have the concept of it at all, must continue
over time, and in most cases must be able to move while still remaining the same thing. Hence movement is pre-eminently that
which makes no real difference. Although other transformations—
everything being reflected into its mirror image, or everything being
doubled in size—are often discussed by philosophers and sometimes
alleged to make no difference, such claims are counter-intuitive,
and, granted the laws of physics, false. But the claim that a mere
displacement in space or time or that a mere reorientation in space
should make no difference carries much more conviction, because
these can be achieved by continuous displacements and reorientations, and therefore are just what are deemed not to make
any difference when they are actual motions of one and the same
physical object. Some principles of relativity are deeply embedded
in our conceptual scheme.

The symmetry of time and space under the group of continuous
displacements is expressed by saying that time and space are HOMO-
GENEOUS, and the symmetry of space under the group of continuous
reorientations is expressed by saying space is ISOTROPIC. In each case
we are saying more than simply that *physical processes* are invariant
under displacement and reorientation: we are saying in addition that
temporal and *spatial features* are themselves unaltered by displacement and reorientation. Although, of course, if I move a thing
to another place, its external relations will naturally be altered, its
internal temporal and spatial relations will not. The period of a
watch tick will be the same tomorrow as it was yesterday: the size
and shape of a car key will be the same in London as in Paris.

It is useful to view homogeneity and isotropy as extreme cases
of more limited symmetries. It might be that there was a natural

periodicity to time; that there was one underlying rhythm we could detect, and know that at each recurrence it would run through the same cycle. We then should have good reason for regarding each period as isochronous, but it might only be a matter of convention how magnitudes were assigned to parts of a period, and although we could be fairly sure that physical processes would be the same if they started at the same "season" of different periods, we could not be so confident if they started at different seasons. But, so far as we know, there is no such natural periodicity. There is no minimum interval, such that processes cannot repeat themselves over shorter intervals, and where, in consequence, we cannot be sure that two physical processes which are otherwise the same but out of phase by some such shorter interval, will develop in the same way. Similarly with space. It could be that there was a fundamental lattice, and it was only displacements through multiples of the unit distance that left everything essentially the same. Or again, it could be that there was a radial symmetry, but only with regard to certain rotations— 90°, like a square, 72°, like a starfish or an apple, or 60° like a honeycomb. But in claiming that time and space are homogeneous we are maintaining that *any* displacement, no matter how large or how small, preserves all temporal and spatial relations, and in claiming that space is isotropic we are maintaining that *any* reorientation, no matter how small, does so too.

The homogeneity of time and the homogeneity and isotropy of space give much support to the view that time and space are themselves unreal, and are not substances at all, but only systems of temporal and spatial relations. It is difficult to state precisely the point at issue in the debate, or unravel the different arguments adduced. It is a paradigm example of a philosophical problem. Almost everyone who thinks about space and time has strong, and often conflicting, intuitions about them; it is difficult to articulate them, difficult to accept anybody else's articulation of them, difficult to determine which arguments bear on which version of the doctrine in dispute, difficult to see how the arguments of others could possibly tell against the version propounded by oneself. Each man must be his own philosopher, seeking to win through to the truth by philosophizing himself, since it is not possible simply to appropriate the philosophical conclusions of others. In this and the next chapter I shall give a partly historical account emphasizing how the various symmetries of space and time have contributed to the problem. But the reader will be dissatisfied, and will want to write out the truth of the matter for himself. He should. It is only by trying to formulate the true view himself and failing to convince others of its truth, that he can become aware of the cogency of the different considerations involved and the complexity of their interplay. At the end of Chapter IX there are further questions which may help those who have already tried to write out their thoughts to write them out a second time in sharper focus.

The debate was begun by Leibniz and Clarke. Clarke was a friend and follower of Newton, and in correspondence with Leibniz sought to defend Newton against a number of Leibniz's criticisms.[1] Leibniz had criticized the doctrine of absolute time and space which Newton had put forward in the Scholium to Definition VIII,[2] and maintained a relative, or better, a relationist,[3] doctrine of time and space. The great success of Newtonian physics led most scientists to suppose that Newton must be right also in his metaphysics, but at the end of the nineteenth century Mach argued strongly for the relationist view on epistemological grounds. Einstein was influenced by Mach, and his Special and General Theories of Relativity have been widely regarded as demonstrating the superiority of the relationist view.

Many other issues are involved. Besides the metaphysical and epistemological arguments of Leibniz, there are other metaphysical, as well as verbal and conceptual, empirical and verificationist arguments for relationism, and mathematical, empirical and metaphysical arguments against. At the most superficial level it is a verbal question, whether Time and Space are properly referred to by substantives, perhaps even dignified by the use of capital letters, or whether they should be seen as merely short-hand expressions for a number of statements like 'Oxford is 80 miles from Cambridge', or 'The battle of Waterloo took place 400 years after the battle of Agincourt', which alone are really capable of being true or being

---

[1] Conveniently available now as *The Leibniz–Clarke* Correspondence, ed. H. G. Alexander, Manchester, 1956; hereafter cited simply as Alexander. All page references are to this edition.

[2] Reprinted in Alexander, pp. 152–60.

[3] The word 'relativity' has now become linked in physics with the two theories put forward by Einstein, the Special Theory of Relativity and the General Theory of Relativity. Although Einstein was led to propound them, in part, by a relationist doctrine of time and space, it is a matter of dispute whether these two theories really are relationist in import. Modern practice, therefore, avoids the word 'relative', and uses the words 'relationist' and 'relational' of doctrines of time and space, and the words 'relativist', 'relativistic' of the physical theories propounded by Einstein.

known. There are deep metaphysical intimations that empty space is a vacuum, a void non-entity, which could not conceivably be any sort of thing, or have any influence on the course of events, and, on the other side, that Time and Space are fundamental categories, of profound importance for all our concepts. There are a welter of different arguments, some empirical, some mathematical, some epistemological, some conceptual, some sound, some fallacious, leading to a number of different conclusions, which we must not assume unthinkingly to amount all to the same thing.

Some relationist arguments are conceptual. It is argued that it is meaningless to refer to spatial or temporal positions except by measuring distance from the axes of some co-ordinate system.[4] It is a difficult claim to sustain because it is our normal habit to refer to times and places in a non-numerical way—'now', 'the year that King Uzziah died', 'Oxford', 'the house on the hill'. It may be that our habits are bad, unscientific ones. But it is difficult to make out that we have all these years been using meaningless locutions unwittingly. Rather than say that it is meaningless, the relationist must fall back on the point already made in Chapter V[5] that it is always possible to replace such locutions by relational ones, and argue furthermore that the ideology of physics requires that everything be characterized in a uniform economical mathematical way, and therefore must discount our habit of referring to dramatic events or familiar places in wordy, qualitative ways. The language of physics is austere, and can accommodate references like AD 1974 or 53 °N, 25 °W, but nothing more personal or vivid. This much granted, he may go on then to claim that quantitative measurements are necessarily assigned to intervals and distances, which are essentially intervals or distances between two points, and therefore are not, properly speaking, to be monadically predicated of one term only, as would be the case if spatio-temporal locations were something absolute, but are, rather, relations between two terms, the frame of reference as well as the

---

[4] For example, by Berkeley, *De Motu*, §§58, 59. Berkeley is arguing against Newton's experiment with globes (see pp. 132–3). "Therefore, if we suppose that everything is annihilated except one globe, it would be impossible to imagine any movement of that globe. Let us imagine two globes and that no other body exists besides. Let us also imagine applying forces in any way: whatever we understand by this application of forces, the circular motion of these globes round their common centre cannot be conceived. But suppose that the heaven of fixed stars is created: a motion will be immediately conceived when the globes approach different parts of this heaven."

[5] pp. 71–2.

spatial or temporal position itself, and therefore necessarily relational. This argument has some appeal, especially when we have in mind only co-ordinate systems of a fairly standard kind. But it is not a water-tight inference from the physicist's stipulation that only mathematical characterizations be allowed. Mathematicians often use real numbers to refer to a point (x, y, z) in an abstract 3-space and only later define a metric by means of a distance function between two arbitrary points (x, y, z) and (x', y', z'). The use of co-ordinates, therefore, does not presuppose a system of measurement. Nevertheless, it suggests one. For it not only assigns to each point in space, or in time, or in space-time, an ordered set of numbers, which serves to identify the point uniquely, but does so systematically.

Besides various topological requirements—that the number of co-ordinates be equal to the dimensionality of the space, that continuity be preserved—we have ordinal ones. If we assign t_1 to one instant t_2 to another and t_3 to a third, and if $t_1 < t_2$ and $t_2 < t_3$, then the instant assigned to t_2 must come between the instants assigned to t_1 and t_3. A co-ordinate system is, as we saw in Chapter VI, a special case of measurement in which the equivalence relation between entities possessing the same magnitude has degenerated into the identity relation, and each entity is, in respect of position, unique;[6] and a measurement system which merely represents ordinal properties is, as we saw, unsatisfactory.[7] We want it to be a genuine scale, not a mere register of degrees, and therefore need to be able to say whether different differences are the same or not. In assigning natural numbers, integers, rational numbers, or real numbers to points we are suggesting that the points have the same structure as the numbers, and that just as the difference between 100 and 99 is the same as the difference between 4 and 3, so the difference between $(100,0,0)$ and $(99,0,0)$ is the same as the difference between $(4,0,0)$ and $(3,0,0)$. It is not an inescapable conclusion. We can define many other distance functions besides the natural one, and can use numbers to label points without imputing to the points any of the metrical properties of the numbers. But to do so is unnatural. The natural way to construe a co-ordinate system is as being a *soi-disant* scale of position, not a mere register of order. The crucial condition for its being a scale is that certain differences of position are given as being equal, and this condition will be satisfied just as well if all the

[6] pp. 83–7.
[7] pp. 91–2.

co-ordinates in any one direction are increased or decreased by the same amount. That is to say the condition for being a scale is satisfied only if the co-ordinates can be transformed by a displacement, without their thereby ceasing to constitute a scale; and hence if we are led to think of a co-ordinate system as being a scale, we require that the space be invariant under displacement. Or to put it another way, being a scale requires the equality of differences, and differences are essentially relations, so that a co-ordinate system does not merely refer to points numerically but emphasizes the importance of the relations that each point bears to others.

If a co-ordinate system is to be used in practice, the importance of relative differences is increased. Although I can conceive of a co-ordinate system in which a difference in the co-ordinates of two points does not correspond in an obvious and uniform way with their difference in position, and although I can talk about it, and write up co-ordinates on the blackboard, and other people can talk about it too, we none of us can use it unless we have some rule for assigning co-ordinates to points and *vice versa*. The rule must be fairly simple: points are many, and life is short. The rule, therefore, must apply in a uniform way, wherever the point is situated. What is the same for different positions is not the positions themselves—they indeed are different—but the differences between different positions. I cannot make out that $(100,0,0)$ is the same as $(99,0,0)$: but I can make out that the difference between $(100,0,0)$ and $(99,0,0)$ is the same as the difference between $(4,0,0)$ and $(3,0,0)$; and if that is so, we can use this sameness that applies in different places to give us a uniform way of actually correlating points and co-ordinates. And so we have a further argument, that space, if it is to be adequately referred to by a co-ordinate system, should be invariant under displacement.

The argument is not conclusive. The standard public procedure for correlating points with co-ordinates does not have to be based on a measurement of intervals, which are necessarily intervals between one point and another, and so can be attributed to one point only relative to the other. Not all measurable magnitudes are intervals. Mass is a measurable magnitude, but is not the measure of the interval between the body to which the mass is being assigned and some arbitrary frame of reference. Mass—rest-mass, that is, to avoid relativistic complications—is an absolute property of things. Although, of course, our assignment of a measure to a mass is, in part, a matter of convention, inasmuch as the numerical measure assigned is relative to the system of measurement employed—the Imperial Weights and Measures, *le Système International*, the mass of the proton—nevertheless our concept of mass, although quantitative, is not in any significant sense relational. Some such quantity could provide a means of referring to spatial positions or temporal instants. It might be the case, for example, that not all periodic processes marked off compatible isochronous intervals, but that some—say the vibrations of the caesium atom—increased in frequency, compared with others, strictly monotonically with the passage of time. We could then date events by giving the frequency of caesium emissions at the time. The carbon dating used by archaeologists is of this form. It offers an entirely mathematical but completely non-relational assignment of temporal co-ordinates.

It is not absolutely inherent in a co-ordinate system that it should be relational. The reason why we do not think of there being a natural zero of position, as there is a natural zero of mass, is because we shape our concept of space so as to be homogeneous and isotropic for other reasons, not simply because it is mathematical. Indeed, in the General Theory of Relativity, each spatio-temporal position has a characterization which, although couched in mathematical terms, is very rich, and might be sufficiently specific to individuate it. Its differential geometry and intrinsic curvature might be as characteristic as that of a hill or a valley is in ordinary life, and equally adequate for securing unique and unambiguous reference. It is not just our requirement of physically significant features being couched in mathematical language that imports a relationist air into the physicist's concepts of space and time, but rather our use of co-ordinate systems subject to certain symmetry conditions. There is no absolute zero in space because we require that there should be none, believing, as we do, that one place is as good as another, and that a difference of position cannot *per se* explain any phenomenon or be of any physical significance. As we saw earlier,[8] this is as much a *fiat* as a fact: whenever there are differences, we ascribe them to other factors, postulating a magnetic field or something else, rather than allowing that a mere difference of position could by itself account for a difference of phenomena. As we become more sophisticated we precisify this vague requirement of the causal inefficacy of space by stipulating that physically significant features of co-ordinate

[8] Ch. V, p. 70.

systems should be invariant under displacement and reorientation.

Leibniz argued that Newton's belief in absolute space was untenable. His own arguments are based on the Principle of Sufficient Reason and the Identity of Indiscernibles. Neither is convincing as it stands. According to the Principle of Sufficient Reason everything must have an explanation, and therefore there would have to be an explanation of why the universe as a whole occupied its actual position in space and time rather than some other one that differed from it only by a reflection or a reorientation or a spatial or temporal displacement:

> 'tis impossible there should be a reason, why God, preserving the same situation of bodies among themselves, should have placed them in space after one certain particular manner, and not otherwise; why everything was not placed quite the contrary way, for instance, by changing East into West. . . . The case is the same with respect to time. Supposing any one should ask, why God did not create every thing a year sooner; and the same person should infer from thence, that God has done something, concerning which 'tis not possible there should be a reason, why he did it so, and not otherwise: the answer is, that his inference would be right, if time was any thing distinct from things existing in time.[9]

The Principle of Sufficient Reason is in part a metaphysical and in part a theological principle. The metaphysical principle is that everything has an explanation. For Leibniz's argument to work he needs to understand this as meaning that everything has a *complete* explanation. It is not enough to give some explanation of why something of this sort occurred: what is required is a full explanation of exactly why this happened as and when and where it did. We must be able to answer every question about it, and say not merely how it was possible that it should happen but why it was necessary that it should have happened in exactly the way it actually did. No question may be left unanswered. It would run counter to Leibniz's principle to claim that any question was unanswerable, or to give the non-answer that it was simply a brute empirical fact that things just happened to be the way they were. There is no room for contingency if everything has a sufficient reason. Everything must be as it is, and a scientist who sufficiently understood the explanations of things would have no need of experiment or observation to discover how things actually were. Thus the Principle of Sufficient Reason in its

[9] Leibniz, III §§5, 6; pp. 26–7. See also Leibniz II §1, p. 16; III §§7–8, pp. 27–8; IV §§1–3, 13–20, pp. 36, 38–9; V §§1–20, 66–73, 76–7, pp. 55–60, 78–81.

strong form runs counter to the fundamental tenet of empiricism, which maintains not only that it is logically possible that things should be other than they are, but that there is some real contingency in the world.[10] Science, and particularly physics, does not seek to explain everything. It discovers regularities, correlations, constant conjunctions, symmetries, patterns, or functional dependences in natural phenomena, but these obtain between certain initial and final conditions, or, more generally, among boundary conditions. But it does not explain the boundary conditions themselves, save in terms of other boundary conditions.[11]

Leibniz argues for the Principle of Sufficient Reason on theological grounds. God is an agent. Agents act for reasons. Therefore God acts for reasons. Ordinary finite agents may, perhaps, sometimes act arbitrarily, but this is a failure of rationality. God must always act for reasons, always for the best of reasons. "God's perfection requires that all his actions should be agreeable to his wisdom: and that it may not be said of him that he has acted without reason; or even that he has prefer'd a weaker reason before a stronger."[12] From which Leibniz infers "When two things which cannot both be together are equally good: and neither in themselves, nor in combination with other things, has the one advantage over the other; God will produce neither of them."[13] We can have some sympathy with Leibniz's doctrine of God, if we see it as a reaction against the view of the voluntarists in the late Middle Ages and of the Calvinists that God was altogether arbitrary in his decisions and the rationale of his decrees utterly inscrutable. An altogether arbitrary God is unintelligible, and we should have no warrant for regarding such a being as a personal agent. But it does not follow that there can be no alternatives open to God which are morally or rationally indifferent, or that it would be a derogation of God's perfection to make an arbitrary choice between them. Indeed, it is consonant with God's having created men as free agents in his own image that he would have made the universe so as to afford some indifferent courses of action in order to give men room to exercise freedom of choice; and while Leibniz's doctrine of God cannot be ruled out of court, it seems on this point at least to be seriously at variance with

---

[10] See above, Ch. I, pp. 1–3.

[11] See E. P. Wigner, *Symmetries and Reflections*, Ch. 4, pp. 38–42.

[12] Leibniz, V §19, p. 60.

[13] Leibniz, IV §19, p. 39.

most men's understanding of God, and open to many objections from Christian theology.

Clarke accepts the Principle of Sufficient Reason, but not in its extreme form; everything can be explained, but not completely. An arbitrary choice by God to create the universe at one time rather than another, or to place it in one position rather than another, would constitute sufficient reason for its existing at that time or in that place.[14] It is difficult not to side with Clarke against Leibniz on the nature of rational choice. When faced by alternatives between which there is nothing to choose, we think it is rational to make an arbitrary choice rather than none at all. Mathematicians often make arbitrary choices of instances, origin, or axes, acknowledging that they could have equally well chosen some other one, and having as the only reason for choosing the one they did the need to make a choice for the sake of definiteness. These considerations are not conclusive against Leibniz. It is possible to conceive of a world in which there are no alternatives that are absolutely indifferent, and no agents making absolutely arbitrary choices. But such a world needs arguing for—it is not forced on us by some necessity of thought nor by an analysis of rational agency, as Leibniz makes out. Although we may be disposed to accept the Principle of Sufficient Reason in some restricted form, in its extreme form it runs counter to other judgements we are led to adopt. Our view of nature is one in which there is much contingency and many things happen by chance and for no sufficient reason: and few theologians would be happy with a concept of God, doomed like Buridan's ass to death or inanition by reason of the impossibility of making up one's mind and arbitrarily choosing just one of two equally attractive alternatives.

Whereas the Principle of Sufficient Reason limits, for Leibniz, the universe on the Godward side, the Identity of Indiscernibles limits it on the manward side. The one bases itself on the limits of divine action: the other on the limits of human knowledge. According to the one, absolute space and absolute time cannot exist because God could have no reason for assigning to anything one spatial or temporal position rather than another: according to the other, absolute space and absolute time cannot exist because men could have no way of telling one absolute spatial or temporal position from another. In

---

[14] Clarke, II §1, p. 20; III §§2, 5–7, pp. 30, 32–3; IV §§1–4, 15–19, pp. 45–6, 49–50; V §§1–20, 66–70, pp. 97–9, 107–8.

part, the Identity of Indiscernibles is a special case of the Verification Principle later put forward by the positivists. The positivists ask 'How do we know whether we are at rest or in uniform motion in absolute space?' and on discovering that we cannot know, lose interest in the question, and the concept of absolute space in terms of which the question is framed. They ask how we could tell whether everything in the universe had been created a year sooner, or whether East and West had been interchanged, and once it appears that it would make no difference to any possible experiment or observation, deny the question any empirical content or scientific interest. Leibniz sometimes, however, seems to be putting forward a stronger thesis, an ontological principle about what can exist rather than an epistemological one about what can be known or what can be said. He is not simply saying that there is no point in a scientist talking about absolute space, but that "there is no such thing as two individuals indiscernible from each other",[15] so that, quite apart from the Principle of Sufficient Reason, it would be impossible for God to displace the whole universe, because to do so would be *agendo nihil agere*,[16] that is, would not be anything at all.

Leibniz hesitates to say that it is logically impossible for two individuals to be indiscernible from each other—although in modern formal logic Leibniz's Law defines identity on just that supposition. But if it is not logically impossible, but only "contrary to the divine wisdom",[17] it is only a consequence of the Principle of Sufficient Reason, and, if the latter is not valid, is not itself established as being valid either. In our ordinary thinking we envisage the possibility of there being two bodies indiscernible from each other, and differing *solo numero*, only because they are two.[18] It could be argued that we deceive ourselves, smuggling in some assumption of a spatial or a temporal difference just in order to differentiate between them and make them numerically distinct: but that argument can be turned on its head, and used to give, as a conceptual justification of space and time, that they are needed in order to enable us to frame the concept of two things qualitatively identical but numerically distinct.[19] It can also be argued that the development of modern physics

[15] Leibniz, IV §4, p. 36.
[16] Leibniz, IV §13, p. 38; cf. V §29, p. 63.
[17] Leibniz, V §25, 26, p. 62.
[18] Leibniz, V §25, 26, p. 62.
[19] See J. R. Lucas, *A Treatise on Time and Space*, London, 1973, §25.

has lent some—but two-edged—support to Leibniz. Different fundamental particles obey various different statistics, which differ from the statistics we use in ordinary life. In ordinary life it is one case if I have telephone number 1895 and you have 2257, and a different case if I have 2257 and you 1895. In quantum mechanics, however, if we are dealing, for example, with photons, we have to use Bose-Einstein statistics in order to calculate results that are borne out by observation, for photons are so indiscernible that we do not count as different two cases where one photon has one frequency and another nearby photon has another frequency, and *vice versa*, but count it as only one. In another way, however, Bose-Einstein particles run counter to the Identity of Indiscernibles, for we can have *two* Bose-Einstein particles which although exactly the same in all respects and entirely indistinguishable from each other are nevertheless two, not one. In a very different way we see the Identity of Indiscernibles at work in mathematics and mathematical physics, when we form an equivalence class, and then take that as the real object of mathematical or physical concern. Thus when we are defining a natural number or a fraction, we form an equivalence class of equinumerous sets or of ordered pairs of natural numbers, and frame our definition in terms of the equivalence class. We show that under cross-multiplication $1/2 \approx 2/4 \approx 3/6 \approx 4/8 \approx 5/10$ and then define $1/2$ as the class of all these, just as we define a particular weight as a class of material objects balancing against one another.[20] On this interpretation, the Identity of Indiscernibles is a *fiat* laying down what shall exist as being of interest to the mathematician or physicist. It is then telling physicists that they should regard as significant not a particular frame of reference with a particular origin and particular axes, but a whole class of frames of reference—all those equivalent under the proper Euclidean group, indeed all those under the Galilean group, of transformations.

Although these interpretations lend some plausibility to some ontological versions of the Identity of Indiscernibles, they do not make it strong enough to rule out the possibility of absolute space and absolute time. That there should be two things perfectly indiscernible from each other but numerically distinct is not logically

---

[20] Compare Leibniz, V §47, p. 71. "I have here done much like Euclid, who not being able to make his readers well understand what *ratio* is absolutely in the sense of geometricians, defines what are the *same ratios*. Thus, in like manner, in order to explain what *place* is, I have been content to define what is the *same place*."

impossible—indeed, could not be, if we are to have a concept of perfect symmetry. If we can conceive of a perfectly symmetrical figure, say a hexagon, then we are claiming that *if* it is turned through 60°, the resulting figure will be identical with the present figure in all respects. It will be qualitatively identical. But it will also be distinct. It will have been turned through 60°, which is a different operation from the identity operation of not turning it through any angle at all. If the Identity of Indiscernibles were a logically necessary truth, then we could have no concept of perfect symmetry. But, whether or not there are any perfectly symmetrical bodies, we do have the concept. If we did not, the conclusions we should be forced to would be that space and time were not homogeneous and that space was not isotropic, not that absolute space and time did not exist. Leibniz's argument against absolute space and absolute time depends on the symmetry groups of displacement and reorientation, and therefore is incompatible with taking the Identity of Indiscernibles as a logically necessary truth. If it is not a logically necessary truth, however, it is not strong enough to rule out the logical possibility of a coherent doctrine of homogeneous and isotropic absolute space and homogeneous absolute time. We may be forced by empirical evidence to reject them, or we may be led to abandon them on other conceptual grounds, and to accept the Identity of Indiscernibles instead. But a strong case needs to be made out, in view of the intuitive plausiblity of Newtonian conceptions of space and time, and their considerable success in providing a schema for referring to and characterizing natural phenomena.

The Identity of Indiscernibles is more plausible in its epistemological form. Newton and Clarke did not take much account of it as a logical or metaphysical thesis, but were sensitive to criticisms of absolute space and absolute time on the score of there being no way of knowing what the absolute frame of reference was. They claimed that there were empirical differences between the absolute and the relational theories of space, and that experiment showed the former to be correct. If a bucket of water is hung by a twisted cord, first the bucket and then the water will spin round. The surface of the water will become concave, as a result of centrifugal force which depends on the water's absolute, not its relative, angular velocity. Centrifugal force could also be detected in a cord connecting two globes which were revolving around their common centre of gravity, and would again be evidence for their having an absolute, and not merely a relative, angular velocity.[21] Newton's bucket experiment and globe experiment showed that at least angular velocity was not merely relative and could be detected by physical experiment; and Clarke repeatedly makes the same point about acceleration. People sometimes accuse Newton of being confused, and mistaking the point at issue. But this is to mistake the logic of Newton's position. The bucket experiment does not prove that absolute space exists—only a few pages earlier he had conceded that there were no physical phenomena to distinguish it from relational space—but, rather, it counters the claim that space must, of conceptual necessity, be relational. If space must be relational, so must velocity, and so must acceleration and also angular velocity. But we can tell whether a frame of reference is revolving or not by seeing whether there are centrifugal forces or not, and if the universe were subjected to an acceleration—a sudden shock, as Clarke puts it[22]—we all should know it. Hence it makes sense to say that a frame of reference is unaccelerated and does not possess an absolute angular velocity because we can in fact discover on occasion by experiment that a frame is or is not accelerating or revolving. Hence it makes sense to talk of a frame which is in a state of rest—has no velocity in absolute space—even though, as Newton admits, there is no physical way of discovering whether a frame is in a state of rest or whether it is in a state of uniform motion; and hence, too, we could also intelligibly talk of some canonical frame of reference, with its axes all in the right direction and its origin rightly situated—perhaps at the centre of things—by reference to which we could give each place its true co-ordinates and characterize its real position correctly.

Newton may still be criticized. Verificationists can still maintain that absolute position, absolute orientation and absolute rest, being undiscoverable are therefore meaningless, while allowing that absolute acceleration and absolute angular velocity, having physical effects, are meaningful concepts. But such an objection is difficult to sustain except on very general philosophical grounds. Newton has conceded all the physicist can ask, and only extreme verificationists will maintain that physics constitutes the whole of knowledge. Newton thought of space as the sensorium of God. God knew where each thing was, not by virtue of its distance from other things but because he had willed that it should be there. God knew, and might

---

[21] Newton, Scholium to Definition VIII; reprinted in Alexander, pp. 157-60.
[22] Clarke III, §4; IV, §13.

communicate his knowledge to man by some special revelation. If Newton by careful study of the Book of Daniel could discover the canonical way of characterizing space, and could tell us which systems were truly at rest and which were merely moving uniformly in a right line, we could understand what he meant, and might well be disposed to accept his word for it. Modern physicists too, while allowing that there were many inertial frames of reference, might have reason to believe that some one of them was to be preferred to all the others. Advocates of the big bang theory of creation are naturally disposed to take the dawn of creation as the natural origin for their time-scale; and, as we shall see, there are other considerations we could imagine which would incline physicists to reinstate some doctrine of absolute space in a Newtonian sense. Verificationist criticisms may persuade us that the concept of absolute space is, as it happens, an idle one and of no significance in physics, but can hardly rule it out as conceptually impossible.

A second line of criticism is to construct a geometry which admits just those entities for which there is good scientific evidence, but no others. This can be done.[23] Instead of taking a traditional concept of space and time, and then factoring out those inertial frames which are equivalent to one another under Galilean transformations, as Newton in effect does, we can define a more complicated and condensed sort of space in which no factoring out needs to be done. It is an alternative approach. It has some verificationist advantages, some conceptual disadvantages. Maybe if Newton were alive today, he would have preferred it. But he would not be obviously wrong if he did not, and would anyhow be much more concerned with the deeper questions about the interrelation of space and time raised by the Special Theory of Relativity.

A third line of criticism of Newton's interpretation of his experiments with buckets and revolving spheres is more thoroughgoing. Mach argued that all Newton had shown was that phenomena were different in different circumstances, but not that this difference should be attributed to an absolute angular velocity rather than to the influence of the other term in the relative angular velocity, namely the mass of the distant stars. We can at the end of the bucket experiment say either that the water is revolving with respect to the

---

[23] L. Sklar, *Space, Time and Spacetime*, Berkeley, Calif., 1974, pp. 202–6; H. Stein, "Newtonian Space-Time", *Texas Quarterly*, 10, 1967, pp. 174–200, esp. pp. 174–84, 194–8.

stars or that the stars are revolving with respect to the water. Angular velocity being essentially relative, it does not matter, on relationist principles, which we say. If the stars revolve round the water in the bucket, they will "draw" the water to the sides, just as much as if the water is revolving with respect to the stars it will show a centrifugal tendency. To quote:

> Newton's experiment with the rotating vesel of water simply informs us, that the relative rotation of the water with respect to the sides of the vessel produces *no* noticeable centrifugal forces, but that such forces *are* produced by its relative rotation with respect to the mass of the earth and the other celestial bodies. No one is competent to say how the experiment would turn out if the sides of the vessel increased in thickness and mass till they were ultimately several leagues thick. The one experiment only lies before us, and our business is to bring it into accord with the other facts known to us, and not with the arbitrary fictions of our imagination.[24]

Mach's argument is at first sight implausible. Although there is no contradiction in adopting a frame of reference according to which the water is at rest and the stars revolving, there are weighty considerations against it. In the first place it seems much more appropriate to assign the angular velocity to water rather than to the stars because it is the bucket that Newton is experimenting with, not the fixed stars. It is easy to explain why the bucket should be revolving, difficult to credit Newton with power to put all the heavens in orbit. Granted some independent knowledge of causes and forces, we have grounds for attributing motion to one body rather than another, and to say that the body on which forces have acted or are acting is the one that is really being moved, and the movement of other bodies with respect to it is not real but only relative. Even Leibniz allows

> there is a difference between an absolute true motion of a body, and a mere relative change of its situation with respect to another body. For when the immediate cause of the change is in the body, that body is truly in motion; and then the situation of other bodies, with respect to it, will be changed consequently, though the cause of that change be not in them.[25]

---

[24] E. Mach, *Science of Mechanics*, tr. T. J. McCormack (Open Court Publishing Co.), 1960, p. 284.
[25] V, §53, p. 74.

We can explain the curvature of the water easily and economically by ascribing it to centrifugal force, whereas it is an extravagant and implausible suggestion that it should be attributed to the stars revolving. Moreover, in the case of simple experiments the experimenter can clinch the matter by making adventitious alterations in the situation and seeing what happens. Since it rests on our arbitrary decisions whether the bucket is revolving or not, we can convince ourselves that the phenomenon is due to what we do to the bucket, not to any influence of the stars.

In the second place it runs counter to deep-seated presumptions of scientific enquiry to attribute phenomena to remote causes rather than to ones near at hand. Although we cannot entirely exclude the possibility of sunspots affecting the trade cycle in China, we look for causes nearer home first. Centrifugal forces might conceivably be due to the fixed stars. But if an alternative account is possible, in which the concave surface of the water is attributed to the movement of the water itself, it is much to be preferred. Mach's alternative explanation is a less attractive alternative explanation just because it explains a terrestrial phenomenon by reference to celestial objects, and although astrology, as a logically possible discipline, cannot be ruled out, there is, as we have seen,[26] a presumption in favour of mundane explanations where possible. A third objection is more sophisticated. According to the Special Theory of Relativity no material object can have a velocity greater than the speed of light. But in having the stars revolve, rather than the water in the bucket, Mach is ascribing to them velocities far greater than the speed of light, and is therefore offering a less felicitous account of the phenomenon.

These commonsensical considerations are not quite decisive. Although they give us good reason for preferring the descriptions 'the bucket is revolving with respect to the stars' to 'the stars are revolving with respect to the bucket', they do not prove that we should say 'the bucket is revolving absolutely' rather than 'the bucket is revolving with respect to the stars'. But equally they do not prove that we should not. Mach and Einstein say we should not, but their reasons are not clear, and seem to shift from the conceptual to the empirical and back again. Mach asks a rhetorical question: "Can we fix Newton's bucket of water, rotate the fixed stars and *then* prove the absence of centrifugal force? The experiment is impossible,

[26] See above, Ch. III, p. 41, Ch. IV, pp. 57–8.

the idea is meaningless, for the two cases are not, in sense-perception, distinguishable from each other. I accordingly regard these two cases as *the same case* and Newton's distinction as an illusion."[27] But there is a great difference between an experiment's being impossible and an idea's being meaningless. To avoid difficulties about the speed of light, let us slightly alter the thought-experiment, and consider the stars being all annihilated. If Newton is right the bucket experiment can be repeated just the same with the same results: if Mach is right, the annihilation of the stars will result in the water's no longer becoming concave. There is a clear empirical difference between the two theories, even if the experiment is impossible to perform. Imagine a Highland lad and lass dancing part of a Scottish reel, and suddenly their heartfelt wish that she should be the only girl and he the only boy in all the world is so much granted that not only all other people, but every other material object whatsoever is suddenly annihilated, as they swing round clasping each other's waist. If Newton is right, the annihilation of everything else will make no difference to their dynamics, and as they rotate they will continue to experience a centrifugal force which only their firm grasp of each other is able to resist. But if Mach is right, with the fading out of existence of all other material objects will vanish also the centrifugal effect, and the two lovers will find themselves not only alone in the universe but falling unresistingly into each other's arms. Other, duller experiments might be more feasible. If we had a very thick bucket, then, in Mach's supposition, its rotation would have an effect on the water even though the water were still. True, it would be small compared to the mass of the stars, but not necessarily so small as to be undetectable. Mach could be proved right by experimental observations, and so, equally, could be proved wrong.

And yet it is misleading simply to dismiss Mach as an unempirical empiricist, who tries to buttress the inadequacy of experiments by claims of conceptual meaninglessness. Although his arguments are fallacious, he is putting forward intimations about the relation of causality with space and time which are of great importance for our understanding of physics. Mach challenges our assumptions about causality and explanation, about non-remoteness and about the constraints we normally impose on possible descriptions of phenomena, in order to maintain a strong doctrine of the causal inefficacy of space and time, to put forward the view that the universe is a unity

---

[27] Quoted Alexander, pp. xlix–l; also *op. cit.*, p. 279.

in which the whole constitutes the environment that makes up the physical properties of each part, and to propose a policy of not restricting the range of possible descriptions but rather seeking formulations of natural laws which are covariant with all possible descriptions.

In Newtonian mechanics we identify "the cause of change" with force.[28] We have an idea of what force is, because we use our muscles to exert and resist forces. We make the bucket spin by exerting a torque on it, and feel the strain in our arms as we dance the eightsome reel. But force so understood is deeply anthropomorphic and some physicists wish to purge physics of all anthropomorphic concepts. For them the only way of knowing that a force is operating is by observing the acceleration it causes. And in order to detect accelerations, we need to know what unaccelerated motions are. If we are to use "the cause of change", as Leibniz was willing to, in order to distinguish real, from merely relative, motion, we need some criterion of what is to count as a change. In the Middle Ages, movement in a straight line was thought to need a force to cause it: to see uniform motion in a straight line as a natural state needing no explanation was a great advance, with a corresponding alteration in where causes should be looked for. If we stick with Newtonian mechanics, we have a moderately simple concept of what does not need a cause to explain it—*viz.* uniform motion in a straight line— at the cost of having to posit centrifugal forces due to absolute space and gravitational forces. One of Mach's insights was that by changing our concept of the normal, that which needs no explanation, we change our concept of cause: and in the General Theory of Relativity, by giving a different, non-Euclidean account of space-time, and by taking geodesics, rather than what we think of as straight lines, for the normal path of material bodies, Einstein was able to give an account of the universe in which no force of gravity had to be postulated, and gravitational and inertial phenomena were seen as both being due to the geometry of space-time.

We should see in Mach's thesis the same interplay of fact and *fiat* as in the case of spatio-temporal location and orientation. Although, as I have argued, we can operate with concepts of absolute position and orientation, we do not want to. We have certain doctrines of the causal inefficacy of space and time, which we attempt to shield

[28] Newton's first law of motion (Alexander p. 160) makes this explicit. "Every body continues in its state of rest, or uniform motion in a right (i.e. straight) line, unless it is compelled to change that state by forces impressed thereon."

from refutation at the hands of awkward facts by postulating further factors—weather, magnetic fields—to save the phenomena. Not all directions are the same for magnetic needles on the earth's surface, but we save the isotropy of space by ascribing the alignment of the needle to a magnetic field which is something different from space. We make space homogeneous by denying physical significance to absolute position or absolute orientation, and insisting that physical phenomena should be invariant under change of position or change of orientation *per se*, and attributing any phenomena that are observed to some other factor. In a similar spirit, Mach and Einstein are unwilling to attribute centrifugal forces to absolute angular velocities, and to find some other factor—the mass of the distant stars—to account for centrifugal forces. They could be right. Even if they were wrong about the stars, there might be some other factor which provided an explanation of why the water was concave when revolving without ascribing it to absolute angular velocity *per se*. But, of course, they also could be wrong. Newton might be right, in that although there was no physical phenomenon which would differentiate one place from another, one time from another, one direction from another, or one velocity from another, there were physical effects which enabled us to differentiate between those that were revolving and those that were not.

So far as Mach's thesis is an empirical question, it has not yet been decided: so far as it expresses an ideal of physical explanation, the issue is not yet clear—the General Theory of Relativity pays a high price for its explanatory successes. On one point, however, Newton was clearly in the right. He has shown that a mechanics which is relational with respect to position, time, orientation and velocity, but not with respect to acceleration or angular velocity, is possible. Newtonian mechanics is just such a mechanics. There is therefore nothing unintelligible about absolute acceleration or absolute angular velocity, and therefore nothing unintelligible about absolute space, absolute orientation or absolute time either. Nor does this depend on the actual results of the experiments. Whatever the form of mechanics, comparable considerations could be adduced. Newtonian mechanics, as it happens, is a second-order mechanics: it is expressed, for constant masses, by the second-order differential equation

$$m\frac{\mathrm{d}^2x}{\mathrm{d}t^2} = f.$$

It is possible to conceive a third-order mechanics in which the corresponding equation is

$$m\frac{d^3x}{dt^3} = f$$

and the equivalent of Newton's first law reads: every body continues in a state of rest, or of uniform motion in a straight line, or of uniform acceleration in a straight line, unless it is compelled to change that state by forces impressed thereon. In that mechanics the initial conditions we needed to know in order to calculate the subsequent development of a system would include not only the initial positions and velocities of each particle, but the initial accelerations as well. Angular velocity would be of no physical significance in that system, but angular acceleration would. If third-order mechanics were correct, Newton could not use his bucket experiment to argue for absolute space, but he could devise a comparable one in which a bucket of water was given an angular acceleration. We could also have a first-order mechanics. It would express the assumptions implicit in some medieval thought. The equivalent to Newton's second law is the first-order differential equation

$$m\frac{dx}{dt} = f$$

and its first law of motion would read: every body continues in a state of rest unless compelled to change that state by forces impressed thereon. In this mechanics the initial conditions would consist only of initial positions, and we could tell by physical experiment whether or not a body, or a frame of reference, was absolutely at rest. Space and time, however, would be homogeneous. There could be alternative versions of first-order mechanics in which space was, or was not isotropic. If it was not, we could also tell by physical experiment how a body was aligned absolutely. In either version no explanation is required for a body's being at rest, but explanation is called for if a body is moving. It is clear that we could go higher or lower. A fourth-order mechanics is conceivable, or that of any finite natural number. At each stage the number of initial conditions would increase, as also the number of features which did not call for, and were not susceptible of, physical explanation. At the limit, we should have an infinite-order mechanics, where every initial condition— position, velocity, acceleration, change of acceleration . . .—had to be given, and nothing was explained. Explanation would vanish, and description take over. Equally, if we went lower, and had a zero-order mechanics, in which spatial co-ordinates were given by simple equations, we should have a system in which each thing had its proper place, and no explanation was called for if it was in its proper place, but some disturbing force would be looked for to account for its not being in its place. No initial conditions would be required in order to work out the natural development of a system, and the explanation would be correspondingly more complete and less contingent. In order to say why a body was where it was, I should not have to refer back to where it was and what its velocity was, at some previous time, say yesterday. But although more complete, explanations would be more difficult to come by. We could not repeat experiments in other places with any warrant for expecting the same results, if each place was different. Indeed, if we adopted a corresponding view of time, we could not repeat experiments at all. If every spatial and temporal co-ordinate could be explained, every spatial and temporal co-ordinate would be relevant, and we could never carry through any experimental procedure to isolate and detect relevant factors. Every factor would be relevant, every instance unique. Our world would consist of qualitatively different individuals—Leibnizian monads in effect—and natural science would be impossible.

It thus emerges that Newtonian mechanics is one among a range of possible mechanics which purport to explain the positions of things. For any finite non-zero positive integer n, there is an nth order mechanics, in which the laws of physics are invariant under translations of motions up to the $(n-1)$th degree, but in which changes of the nth degree are detectable by physical experiment and call for a physical explanation. We shall be able to view motions up to the $(n-1)$th degree as relational, but be impelled to regard these of the nth degree as absolute.[29] Hence, if we are to have a mechanics which gives any sort of scientific explanation, it will generate a symmetry group, in which displacements of space and time, and perhaps some but certainly not all of their derivatives, are causally irrelevant, and are treated as initial conditions. In Newtonian mechanics as it happens, uniform velocity and reorientations are not

---

[29] But see John Earman, "Who's Afraid of Absolute Space?", *Australasian Journal of Philosophy*, 48, 1970, pp. 296–7; and L. Sklar, *Space, Time and Spacetime*, Berkeley, Calif., 1974, III D3, pp. 202–6.

physically detectable, but acceleration and angular velocity are. There will always, therefore, be a verificationist argument against absolute positions, orientations, velocity, . . ., etc. up to the transformations of the symmetry group, countered by bucket arguments showing that since further derivatives are physically significant, and we can by experiment determine for which systems they are absolutely zero, it cannot be conceptually impossible to think of systems having absolute positions, orientations, or velocities.

Further Reading

A   For a careful criticism of Leibniz's relationist views, see
Graham Nerlich, *The Shape of Space*, Cambridge, 1976, Ch. I.

B   For a clear survey of arguments on both sides, see
C. Hooker, "The Relational Doctrines of Space and Time", *British Journal for the Philosophy of Science*, **22**, 1971, pp. 97-130.

See also

A   D. Sciama, *The Unity of the Universe*, London, 1959, Ch. 7, pp. 84-101; Ch. 14.

B   J. C. Graves, *The Conceptual Foundations of Contemporary Relativity Theory*, MIT. Press, Cambridge, Mass, 1971, Ch. 4, §17, pp. 298-305.

B   C. D. Broad, *Scientific Thought*, London, 1923, pp. 161-2.

B   E. Nagel, *The Structure of Science*, London, 1961, Ch. 8, §1, pp. 203-14.

B   C. W. Kilmister, *The Environment in Modern Physics*, London, 1965, Chs. 2 and 3, pp. 9-42.

B   G. Berkeley, *De Motu*, esp. §§58-62.

B   M. Gardner, "Relationism and Relativity", *British Journal for the Philosophy of Science*, 28, 1977, esp. §§1-3, pp. 215-21.

B   L. Sklar, *Space, Time and Spacetime*, Berkeley, Calif., 1974, Ch. III, pp. 157-234.

B   H. Stein, "Newtonian Space-Time", *Texas Quarterly*, 10, 1967, pp. 174-200.

C   Hans Reichenbach, *Space and Time*, New York, 1957, §34.

C   John Earman, "Who's Afraid of Absolute Space", *Australasian Journal of Philosophy*, 48, 1970, pp. 287-319.

Preliminary Reading for Chapter IX
A   *Feynman Lectures*, Vol. I, Ch. 52.
A   Kant, *Selected Pre-Critical Writings*, tr. G. B. Kerferd and D. E. Walford, Manchester, 1968, Ch. II, pp. 36–43.
B   E. P. Wigner, *Symmetries and Reflections*, Bloomington and London, 1967, pp. 57–62, 74–5.

IX

Reflections, Relationism, and Parity

The continuous transformations perplex us because they are so natural: we find it easy to redescribe the phenomena in terms of the movements of material objects without recourse to invisible abstractions like absolute space. The discontinuous transformations perplex us because they are unnatural: reflection, inversion, change of parity lead us from the familiar world into one that seems strange and uncanny, because although it retains many familiar features, so much so as often to give us a strong sense of *déjà vu*, yet it remains altogether inaccessible to us as a matter of practical possibility. We never can get through the looking glass. We can recognise things in the looking-glass world, and if only we could get there could find our way about in it. But we cannot envisage any physical manoeuvre by means of which our left and right could be interchanged, and therefore we realise that we are forever excluded from this the other half of imaginable existence. It is very tantalising.

Kant thought that the phenomenon of incongruous counterparts, as he termed them, provided a stronger argument against Leibniz than any of those adduced by Clarke. Although the bucket experiment, and considerations subsequently put forward by Euler, had convinced him that space was, indeed, absolute, they were only arguments from physics, and something more *a priori* would be more conclusive.

The proof which I am seeking here is intended to place in the hands, not of engineers, as was the intention of Herr *Euler*, but in the hands of geometers themselves a convincing proof that would enable them to assert, with the clearness customary to them, the reality of their absolute spaces.[1]

[1] "Concerning the Ultimate Foundation of the Differentiation of Regions in Space" in Kant, *Selected Pre-Critical Writings*, Manchester, 1968, pp. 37–8.

Mirror images, although in one sense exactly like their originals, are, in another, unlike. So too, a right hand and a left hand are both very much the same, and yet deeply different. Every feature of a right hand is exactly paralleled by a corresponding feature of a left hand, and yet we cannot superimpose the one on the other, no matter how we move it about. If the relationists are right, "space merely consists of the external relations of the parts of matter which exist alongside one another".[2] But the external relations of the parts of a right hand existing alongisde one another are exactly the same as the external relations of the parts of a left hand existing alongside one another. It is easier to picture two-dimensional figures

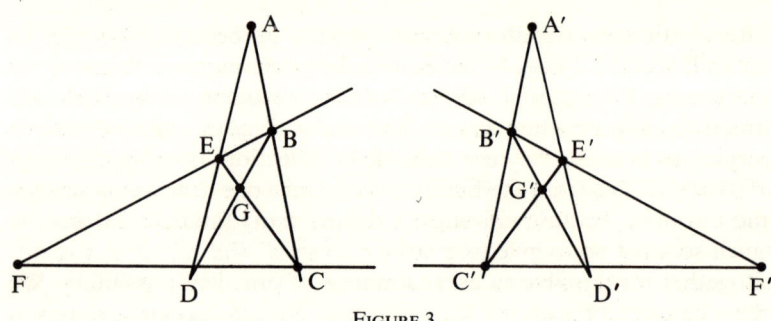

FIGURE 3

ABCDEFG and *A'B'C'D'E'F'G'* which are mirror images (see Figure 3). *AB* = *A'B'*, *AF* = *A'F'*, etc. and ∠*BAF* = ∠*B'A'F'*, ∠*DBC* = ∠*D'B'C'*, etc. It is clear that *ABCDEFG* and *A'B'C'D'E'F'G'* are completely similar to each other and the same size, and yet there is some difference: they are not congruent; it is not possible by any rigid movement in the plane to superimpose one figure on the other. If the relationist claims to give a complete account of spatial properties in terms of the distances between points and the angles between lines, then they are faced with a clear counter-example in there being figures which are identical as regards distances and angles yet none the less different in some other spatial respect.

Kant's argument has been much attacked. Remnant and Bennett attack it on verificationist grounds.[3] Kant considers the possibility

² Kant, *op. cit.*, p. 43.

³ P. Remnant, "Incongruous Counterparts and Absolute Space", *Mind*, 72, 1963, pp. 393–9; J. Bennett, "The Difference between Right and Left", *American Philosophical Quarterly*, 7, 1970, pp. 175–91.

of the first created thing being a hand, and says that it must be either a right hand or a left hand. But how could we tell which? All our normal ways of distinguishing right from left depend either on other asymmetries in the external world or on imagining ourselves, with our asymmetrical bodies or asymmetrical habits of mind, being present in the universe, and so on there being some external point of reference relative to which right is being distinguished from left. Remove all external reference points, and it would be impossible to tell right from left, and Kant's thought-experiment loses empirical content. It would be quite impossible to say whether an isolated hand was a right hand or a left hand if it was the only thing in an otherwise empty universe. More generally, Bennett argues, leaving aside recherché subtleties about the non-conservation of parity, we could not correct a deviant language-user who systematically interchanged the words 'right' and 'left', and related words, except by relation to a common material frame of reference. And so, they maintain, relationism is rehabilitated.

These arguments are like those of Leibniz, and open to the same objection. It is quite possible for there to be a difference without our being able to decide which label to attach to which item. Just as we can see that a regular hexagon is symmetrical although we cannot tell whether it has been turned through 60° or not, so we can see that two figures are mirror images of each other, although we cannot tell which should be regarded as right or left, and so also we can see that one figure on its own would have an incongruous mirror image if it had one at all. Or, to take a very different example from another science, biologists might find that specimens of apparently the same species, instead of being inter-fertile, fell into two exclusive classes, the members of each being inter-fertile with members of the same class but not with members of the other. We could imagine that the closest examination yielded no discoverable difference between the two species, not even in their DNA under the electron microscope. Given any one specimen we could not tell which species it was a specimen of. But we should have good empirical warrant for believing that it must either be a specimen of the one or else be a specimen of the other. The fact that, in the absence of other, labelled specimens, we could not tell which it was, would not at all prevent its being one or the other. Verificationist arguments are crude, and often lead to false conclusions.

The important thing about left and right is not that we should tell which is which, but that we can see that they are different. At the very least, it is enough to refute the simple relationist claims that space is merely a system of spatial relations where spatial relations are taken to be just distances and angles. It can also be taken as positively supporting absolute space, although more argument is needed than Kant gave.[4] There is an evident connexion between the incongruity of counterparts and the dimensionality of space. In saying that the figures *ABCDEFG* and *A'B'C'D'E'F'G'* could not be superimposed one on the other, it was necessary to specify that it was not possible by any rigid movement *in the plane*. It would be perfectly possible to superimpose the figures in a three-dimensional space by revolving one 180° round the axis of reflection. That would constitute a continuous rigid movement by means of which *ABCDEFG*, say, could be transformed into *A'B'C'D'E'F'G'*, but it would involve taking it out of the plane. Similarly with a three-dimensional space, many philosophers have thought that three-dimensionally incongruous counterparts could be made congruous if they could be moved in a fourth dimension.[5] Indeed, it is generally true that in any ordinary n-dimensional space there will be configurations which are counterparts to each other and, in *n*-dimensional space, incongruous, but which are congruous if the n-dimensional space is embedded in an $(n+1)$-dimensional space. The incongruity of counterparts is, thus, connected with the dimensionality of space. But the dimensionality of a space is a property of the space as a whole. It does not depend on there being material objects that a space be two- or three-dimensional, or that it can contain incongruous counterparts. These are topological features of the space, and are naturally expressed in subject-predicate form, with the word 'space' as subject. We are thus led to use the word 'space' as a full-blooded substantive, which constitutes an argument for saying that space is a substance, a bearer of (topological) properties whose consequences we can notice in ordinary experience.

Ordinary n-dimensional spaces admit of configurations which are counterparts of, while not being congruent to, each other. But not all spaces are ordinary. The most familiar example of a non-orientable space, as it is called, is a Möbius strip. A Möbius strip can be made by taking 8 inches of gummed tape, and sticking the *gummed* sides

---

[4] I am much indebted to Graham Nerlich, *The Shape of Space*, Cambridge, 1976, Ch. 2, for providing the needed argument.

[5] L. Wittgenstein, *Tractatus Logico-Philosophicus*, London, 1922, 6.36; Max Jammer, *Concepts of Space*, New York, 1960, pp. 131–2.

at each end to each other. It turns out to have, contrary to expectation, only one surface and one edge, and if *ABCDEFG* and *A'B'C'D'E'F'G'* are counterparts, they can be superimposed by sliding along the strip.[6] Another example is the surface of a sphere with the two points at the opposite ends of each diameter identified. Most spaces are orientable, but some, like the surface of a Möbius strip, are non-orientable, and in a non-orientable space, counterparts can be made congruent much as if an ordinary space is embedded in a higher-dimensional space. We have every reason to think our space is orientable, but we cannot absolutely exclude the possibility that some trip beyond the galaxies or some tangling with a black hole might interchange left and right. But, once again, whether a space is orientable or not is a topological property of the whole space. If our space turns out not to be orientable, then it is a surprising feature, but a feature that is none the less a property of the whole space. Whether counterparts are actually incongruous—as we believe—or not is not a fact about any particular material objects but about space as a whole, and so constitutes a warrant for talking about space in a rather reverential way, as an entity whose properties pervade our experience, and which therefore can be regarded as an important item in the furniture of the world.

There are other topological properties of space as a whole, besides the fundamental one of continuity, and those of dimensionality and orientability. Our space is, we believe, simply connected: there is no way of getting from the inside of a sphere to the outside without going through the surface of a sphere. Once again, this would not be the case if we were to embed our space in a four-dimensional one; and once again we are familiar with two-dimensional spaces, such as the torus or ring, in which a continuous closed curve does not necessarily divide the surface into two disconnected parts. Sklar cites a version of connectedness to argue that there is nothing special about the incongruity of counterparts, and that therefore Kant has done nothing to refute relationism.[7] But the fact that there are other topological features does not detract from the incongruity of counterparts being a striking topological feature which is naturally ascribed to space as a unified whole. Other topological features are

---

[6] For a fuller and clearer exposition, see Graham Nerlich, *The Shape of Space*, Cambridge, 1976, Ch. 2, pp. 36–7; or Graham Nerlich, "Hands, Knees, and Absolute Space", *Journal of Philosophy*, LXX, 1973, pp. 337–51.

[7] L. Sklar, "Incongruous Counterparts, Intrinsic Features, and the Substantiviality of Space", *Journal of Philosophy*, **71**, 1974, pp. 277–90.

more sophisticated: the incongruity of counterparts is accessible to anyone who has been puzzled by a mirror or read *Alice Through the Looking Glass*, and provides a neat *argumentum ad hominem* against the relationist whose doctrine of spatial relations commits him to the significance of features invariant under the Euclidean group of transformations. Nor does Sklar's conclusion follow. A really determined relationist can translate talk about the topological features of space into talk about spatial relations between objects, but at a price, a price which will make relationism appear to many an unattractive option. Since the topological properties of space affect the spatial features of objects, we can expect to characterize the former in terms of the latter; but in doing so we have to invoke principles and complexities that deprive the relationist account of its claim to economy and simplicity. Leibniz had originally maintained that "space denotes, in terms of possibility, an order of things which exist at the same time, considered as existing together",[8] and Sklar says that, according to the relationist view, "space is nothing but the collection of actual and possible spatial relations between actual and possible material objects".[9] The invocation of *possible* objects and *possible* spatial relations, Sklar admits, is essential, as the invocation of possibilities of sensation is essential for the phenomenalist. But collections of possibilities are, to say the least, highly abstract and ethereal entities, open to as much objection on the score of reality as space ever was, and far more numerous. Again, Sklar can, indeed, characterize the property of being right-handed as one that is preserved under all continuous rigid motions in a three-dimensional space, although not preserved under all continuous motions in a four-dimensional space;[10] but to speak of "all continuous rigid motions" in a three- or four-dimensional space is to quantify over a large range of entities, and thereby to confer on them a large measure of ontological respectability. If all continuous rigid motions in a three-dimensional space are real, we have not saved much by denying the reality of the space itself.

Kant was on to something. He was tackling the question, raised in the Leibniz–Clarke correspondence,[11] whether space was a substance, an attribute or a relation, whether it was an entity or a void

---

[8] Leibniz III §4, p. 26; compare V §104, p. 89.

[9] Sklar, p. 281.

[10] p. 285

[11] Leibniz III §§3, 4, pp. 25–6; Clarke III, §3, pp. 31–2; Leibniz IV §§8–12, pp. 37–8; Clarke IV §§7–12; Leibniz V §§36–51, pp. 66–73; Clarke V §§36–46, pp. 103–4.

non-entity, whether it was composite and made of parts or a unity not susceptible of division and having to be considered as a whole. Kant is rejecting the assumption that space must be either a substance or an attribute or a relation. Logically speaking, it is a substance, properly referred to by a singular noun. But although singular, it is not simple. It is an integrated structure, and although we can talk of regions, or parts, of space, we cannot talk of those regions or parts in isolation, but only in the context of the whole. Newtonian space is not just a set of points. It is a set of points, together with a family of open sets such that distinct points are contained in disjoint open sets, continuous, differentiable, three-dimensional, orientable, simply connected, with a definite metric which is Euclidean. Kant's arguments fall far short of establishing this much, and after publishing his paper "Concerning the Ultimate Foundation of the Differentiation of Regions in Space", he came to the conclusion that space and time are not discovered by us as objective features of the world, but are, like causality, imposed by us on the world as a framework for organizing experience. In the *Prolegomena* he cites the problem of incongruous counterparts as an argument against space being real, and in favour of its being the form of outer intuition.[12] The argument is difficult to follow, and relies greatly on the general principles of Kant's critical philosophy.

Although we can read out of Kant's earlier paper an argument against relationism, we should not see it as an argument for absolute space and absolute time as Newton propounded them. Space and time may be entities with a definite structure and definite properties, as Newton thought, but with a different structure and different properties from those Newton supposed. Newton thought of absolute space and absolute time as two distinct absolutes, and that any spatio-temporal characterization of phenomena was to be decomposed in a unique way into a separate spatial and a separate temporal characterization. In the Special and General Theories of Relativity we think not of space and time, but of space-time. This space-time is four-dimensional, but with one of its dimensions being different from the other three—which we express by saying that it has an indefinite metric with Lorentz signature. But this is still to talk about space-time as a whole. And to that extent Kant's original argument stands.

[12] *Prolegomena to any Future Metaphysic*, tr. P. G. Lucas, Manchester, 1953, §13, pp. 41–51.

Although absolute space and absolute time have been replaced by space-time, some differences between space and time, as shown by the indefinite metric with Lorentz signature, remain. Timelike directions are different from spacelike directions, and we may wonder what the effect of temporal reflection might be. Newtonian mechanics and Maxwell's equations for electromagnetism are both invariant under time-reversal. In fact, outside thermodynamics most physical laws are invariant under time-reversal. (But not all. There is, for example, a breakdown of time-reversal invariance in neutral kaon decays.) If a film of a physical process were run backwards, it too should show a possible physical process, and a physicist would be as unable to detect the difference as Leibniz reckoned we should be if East and West were interchanged.[13] We feel very uneasy at this. Time as we know it in all our thinking and doing is directed. The future is quite unlike the past, and the present, which comes between them, has special status. Physics knows nothing of the special status of the present, and, for the most part, nothing of the directedness of time. To this extent physics has a different, and more meagre, concept of time. In forming their concept, physicists have abstracted from our everyday understandings of time leaving out some features we normally regard as essential.[14] Even in our understanding of physical reality, however, we need more than the totally reversible time of Newtonian mechanics. We need the concept of cause, and causes are at least weakly antecedent to their effects.[15] Causes not only regularly precede their effects, but explain them, and therefore need to be in some sense simpler than them, which in turn introduces an asymmetry in information, or negative entropy, as it is sometimes called, between causes and effects.[16] Once we allow that directedness is a fundamental feature of time, we are committed to its having only one dimension. For if there were more than one dimension of time, we could reorient a physical process so as to have it aligned in any one dimension in the opposite sense, and

---

[13] Leibniz III §5, p. 26, quoted above, Ch. VIII, p. 127.

[14] For a fuller account of different concepts of time, see K. A. Denbigh, *Three Concepts of Time*, Berlin, 1981.

[15] See above, Ch. III, p. 33.

[16] See discussion started by K. R. Popper and continued by Richard Schlegel, E. L. Hill, A. Grunbaum, and R. C. Bosworth in *Nature*, CLXXVII, 1956, p. 538, CLXXVIII, 1956, pp. 381-2, CLXXIX, 1957, p. 1297, and CLXXXI, 1958, p. 402. See also G. J. Whitrow, *The Natural Philosophy of Time*, Edinburgh, 1961, Ch. I, pp. 7-10.

directedness would not be a fundamental feature.[17]

Reflection is not the only operation that satisfies the group-theoretical equation

$$R^2 = I.$$

If we do not simply reflect in the plane (or hyperplane) perpendicular to one axis, but invert every co-ordinate, replacing x by $-x$, y by $-y$, z by $-z$, etc., we obtain a change of parity. It is obvious that if the operation is performed twice we are back where we started,

i.e., $P^2 = I.$

In a one-dimensional space $P = R$. In a more-than-one-dimensional space we can obtain a parity operator by successive reflections in different planes. Thus in three dimensions we obtain a parity operator by reflecting first in the YZ-plane (perpendicular to the X-axis), then in the ZX-plane, then in the XY-plane (the order of reflection is immaterial). Parity is of great importance to physicists. For a long time it was thought that parity was conserved, and that the laws of physics would apply exactly the same way under a change of parity as they do now. The discovery that parity was not conserved in "weak decay" was of great importance. But even though parity is not conserved, it seems that if the direction of time (T) were reversed, and positive and negative charge (C) were interchanged as well as parity (P), all physical phenomena would remain the same. Even then there are further possibilities to perplex us. If we came to know of other sentient beings who appear to be more and more like us, we could never tell whether they are really like us or anti-material counterparts, until on first seeing them we see whether they stretch out their right or left hands to greet us![18] We should note that whereas in odd-dimensional spaces a parity operator transforms a configuration into its incongruous counterpart, in even-dimensional spaces this is not so. For example, in a two-dimensional space, the parity operator transforms a figure ABC into $A'B'C'$, which is congruent to it, and could equally well be obtained by a reorientation

[17] For a very different argument, see J. Dorling, "The Dimensionality of Time", *American Journal of Physics*, XXXVIII, 1970, pp. 539–40.

[18] *Feynman Lectures* I §52–8. Martin Gardner, *The Ambidextrous Universe*, 2nd ed., Pelican, 1982, pp. 194–6, 228.

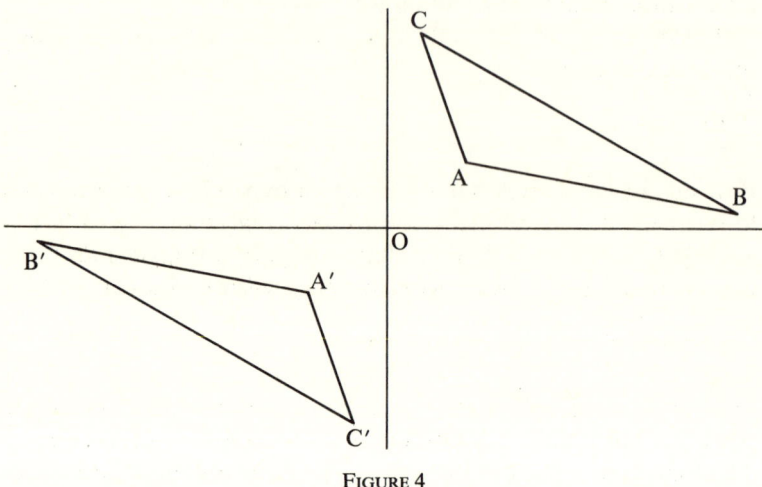

FIGURE 4

through 180° (see figure 4). Hence it is only in odd-dimensional spaces, such as ours, that parity operations involve changes from left-handedness to right-handedness or vice versa, or does anything that could not be accomplished equally well by reorientations. The fact that parity has come to be of great importance in modern physics might be seen as depending on the three- (or at least odd-) dimensionality of space.[19]

From earliest times men have had an intimation of a worrying dualism in the physical universe, as the overtones of the word 'sinister' show. No matter how often we purge our science of this shadow reality, theories expand to re-instate the possibility of it. In part it is due to the group theory underlying our concept of symmetry. Almost all groups have subgroups, and indeed, almost all have subgroups exactly half as large. In all such groups there will be a parity property. The division between those elements that are members of the subgroup and the others will not only be mutually exclusive and jointly exhaustive—that would be true of any subgroup—but is such that every member of the group "commutes with" the subgroup. The subgroup can be regarded as itself a sort

[19] For detailed argument, involving relativistic quantum field theory, see S. P. Rosen, "TCP Invariance and the Dimensionality of Space-Time", *Journal of Mathematical Physics*, **9**, 1968, pp. 1593-4.

of unity, and in this case it does not matter what then the order of operations is. Such a subgroup is termed an invariant subgroup. It is invariant subgroups which generate alternative standards of symmetry. If a subgroup is not invariant, if in general the coset HS is not the same as the coset SH, we shall not attach much importance to the homomorphism mapped by the subgroup onto the original group, as we can alter it by commuting the subgroup and the other element. An invariant subgroup introduces a much firmer break-up into the group structure. Where the subgroup is of index 2 (i.e. half as large as the whole group), every element of the group which is not an element of the subgroup S, must have its coset with respect to S, the only coset available, i.e. that composed of all the elements of the group which are not elements of the subgroup. So its left- and right-cosets must be the same, and the subgroup thus invariant. It is really a corollary of the law of the excluded middle that every subgroup of index 2 is invariant, and those who have seen in parity phenomena a reflection of the fundamental distinction between the positive and the negative approach have not been far wrong.

Further Reading

Arguments for and against relationism are given in

A　　Graham Nerlich, *The Shape of Space*, Cambridge, 1976, Ch. 1, pp. 5–28.

A　　W. H. Newton-Smith, *The Structure of Time*, London, 1980, Ch. 1 §§4, 5, pp. 6–10; Ch. 4, pp. 79–95.

B　　L. Sklar, *Space, Time and Spacetime*, Berkeley, Calif., 1974, pbk. ed., 1977, Ch. III, pp. 157–234, with useful bibliography.

Reflection, incongruous counterparts, parity and time reversal are discussed in:

B　　Lewis Carroll, *Alice Through the Looking Glass.*

B　　John Earman, "Kant, Incongruous Counterparts and the Nature of Space and Space-time", *Ratio*, III, 1971, pp. 1–18.

B　　L. Sklar, "Incongruous Counterparts, Intrinsic Features and the Substantivality of Space", *The Journal of Philosophy*, LXXI, 1974, pp. 277–90.

C　　P. Remnant, "Incongruous Counterparts and Absolute Space", *Mind*, 72, 1963, pp. 393–9.

C　　Jonathan Bennett, "The Difference Between Right and Left", *American Philosophical Quarterly*, 7, 1970, pp. 175–91.

B　　Martin Gardner, *The Ambidextrous Universe*, Basic Books, U.S.A., 1964, Penguin, 1967, and Pelican, 1970; 2nd ed., Pelican, 1982.

C　　D. F. Pears, "The Incongruity of Counterparts", *Mind*, 61, 1952, pp. 78–81.

C　　J. J. C. Smart, *Between Philosophy and Science*, New York, 1968, pp. 217–18.

C　　L. Fonda and G. C. Ghirandi, *Symmetry Principles in Quantum Mechanics*, New York, 1970, esp. Ch. 6.

Exercises

(1) In Chapter VIII, p. 120 the suggestion sometimes put forward by philosophers that if everything doubled in size it would make no difference was said to be false, granted the laws of physics. Is it obviously so?

(2) If the reader has already tried to write out his own views about relationism, he may then find it useful to distinguish different claims made by different brands of relationism, and consider how these distinctions could be further refined. He should also identify the different motives that people have for being and for not being relationists, and the price they are prepared to pay. In so far as it is a metaphysical doctrine, it is illuminating to consider how far it is based on actual empirical evidence, or could be refuted by possible empirical evidence.

B   Bertrand Russell, "On the Notion of Cause", *Proceedings of the Aristotelian Society*, xiii, 1912-13, pp. 1-26; reprinted in *Mysticism and Logic*, London, 1918, pp. 180-208, Pelican, 1953, pp. 171-96.

X

Causal Cords

THE simple concept of a causal law which we elucidated in Chapters III and IV was discrete. Temporally it distinguished two stages, an earlier and a later, the one when the antecedent conditions obtained, the other when the consequences emerged. With respect to each causal factor or condition, it distinguished two possibilities, that the factor was present or that it was absent. Each factor, thirdly, was sharply distinguished from every other one: we had a discrete set of factors A_1, A_2, ... A_n. And, fourthly, we operated with a binary, black-and-white logic, with two discrete truth-values, True and False: a putative causal law was either True or False, and must be one or the other, and could not be anything in between. Now, however, that we have established the principle of continuity, we need to generalise the discrete concept of a causal law to accommodate continuous variations. We can consider continuous variations with respect to

(1) time
(2) causal factors
(3) number of causal factors
(4) truth-values.

The generalisation in respect of the first two, time and causal factors, yields the "functional dependences" of classical physics, normally expressed in terms of differential equations. These will be our concern in this chapter. Generalisations in respect of the third item—number of causal factors—is relevant to field theories in classical physics, and especially important, together with generalisation of truth-values, in quantum mechanics. In quantum mechanics, however, the programme of everywhere replacing discrete by continuous variables runs into difficulties, and discreteness re-emerges as an ineliminable characteristic of Nature.

Hume spoke of constant conjunction, whenever A then also Z. We can view constant conjunction as a special sort of relation, a *function* or a *mapping* or a *transformation*. Any relation can be expressed as a set of ordered pairs. All pairs of a particular type of cause and a particular type of effect are members of the set of causes-and-effects; or in symbols

$$\langle A, Z \rangle \in C.$$

A function or a mapping or a transformation is a many–one relation. Given any particular cause, then there is only one effect that can follow; or in symbols

$$\langle A, Z \rangle \in C \,\&\, \langle A, Z' \rangle \in C \to Z = Z'.$$

We can also think of this as a function holding between event-types, with different event-types, representing different sorts of cause as its domain, and different sorts of effects as its range; or in symbols

$$Z_1 = f(A_1)$$
$$Z_2 = f(A_2) \text{ etc.}$$

A_1, A_2, etc., may well be complex, being fully specified in disjunctive normal form as in Chapter IV.[1] If we accept, as before, the principle of there being only a finite number of possibly relevant causal factors, we can see each fully specified type of cause as a "point" in a "logical space". The logical space has as many dimensions as there are different possibly relevant causal factors. In each dimension there are only two points, one representing the presence of the possibly relevant causal factor, the other its absence—in mathematical terms, it is an n-dimensional vector space over the integers modulo 2. If there are n possibly relevant causal factors, there will be 2^n distinct points in logical space. If we decide that certain factors are irrelevant, we can express this by projecting the original space onto a sub-space spanned by all the relevant factors, but not the irrelevant ones. In particular, if we regard just one factor, say Z, as relevant we can project onto just the Z axis. This is what we have been doing with regard to effects.

[1] See above, pp. 49–50.

In Chapter IV we did not observe the second half of Hume's fourth rule that the same effect never arises but from the same cause.[2] Although from the practical point of view it is sensible not to distinguish arsenical death from cyanide death, there is some difference between the two states of affairs—else forensic scientists would never be able to tell the cause of death—and it is a reasonable idealisation to make it a rule always to distinguish those Z situations which follow upon an A_1 cause from those that follow upon an A_2 cause, etc. It means, essentially, that instead of describing the effect simply as being Z we fill out the description to Z & Y_1 or Z & Y_2, in much the same way as we needed to fill out the description of causal situations; or, more simply, we change our notation, and characterize the effect in terms of possible effect factors, $Z_1, Z_2, \ldots Z_n$, analogously to the possible causal factors $A_1, A_2, \ldots A_n$. In that case the causal relation becomes a one-one mapping of the logical space onto itself. Granted that there are only a finite number of possibly relevant causal factors, we can adduce in favour of Hume's stipulation the consideration that if it were many-one, then a causal system starting from the initial condition would at each causal step lose information and become less fully specified, and after only a finite number of steps reach a minimum in which no further change was possible. If a mapping of a finite space is only a mapping *into* itself, and not a mapping *onto* itself, then the mapping can be iterated only a finite number of times. The generalisation of this argument, if it could be carried through to the infinite case, would give us an analogue of the Second Law of Thermodynamics. If we insist that the causal relation be one-one, we shall avoid problems of information loss, but shall be led to other surprising conclusions, so long as our system is only finite. For if there are only a finite number of points in condition space, say 2^n, then after 2^n steps at most our mapping must bring us to a point which we have been at already, and therefore, granted that it is a one–one mapping, must bring us back to our initial condition. Hence we have a simple ergodic theorem establishing the cyclical nature of change. Hume uses this argument as rebutting one of the proofs of the existence of God,[3] and it may be that some intimation of this argument led philosophers in the ancient world to conclude that time itself was cyclic.

---

[2] p. 44, and p. 67.

[3] *Dialogues Concerning Natural Religion*, part viii, *init.* (Hafner edition, p. 53).

We can generalise the model of discrete causality given in Ch. IV,[4] in which the set of possible causal factors is finite and known, but the factors are not causally independent. Instead of having discrete factors, say, A_1, A_2, . . ., A_n, each of which is either present or absent, we have variables, each taking as its value in a particular case some real number, say, a_1, a_2, . . ., a_n; each of these is to be understood as giving a numerical characterization of some aspect—weight, energy, temperature, wavelength—of the causal situation. The effect, fully specified in accordance with the reasoning of the previous paragraph, is similarly characterized by real numbers, z_1, z_2, . . ., z_n, each of which gives a numerical value for some aspect—weight, energy, temperature, wavelength—of the effect. Instead of saying, as in the discrete case, that the presence or absence of Z depends on the presence or absence of some combination of A_1, A_2, . . ., A_n, we say that the numerical value of each of z_1, z_2, . . . z_n, depends on the numerical value of a_1, a_2, . . ., a_n. Each z is a function, different for each different z, of a_1, a_n, . . ., a_n. Given n functions, one for each of z_1, z_2, . . ., z_n, we can calculate for any causal situation, a_1, a_2, . . . a_n, what its effect, with respect to each of the same n aspects will be. We have replaced a simple cause-effect relation by a functional dependence.

It is natural to go a stage further, and generalise not only from discrete causal factors but from the articulation of events into antecedent causes and subsequent effects. Instead of considering just two *given* times, one when the causes start to operate, the other when the effects eventuate, we consider how the state of the system at any one time is correlated with its state at *any* other time. We make an arbitrary choice of which conditions to consider as initial conditions, and then consider what conditions will obtain after the elapse of some arbitrary interval, δt: causal chains have given way to causal cords.

The concept of a causal cord has two advantages. First, it releases us from the illusion that the cause must be unique in time. By this I mean that each event has just one cause a certain time before it and no other circumstance before or after that one cause can also be the cause of the event. Where there is a natural articulation of situations, there is one privileged cause: but where there is none, there are many causes preceding and succeeding one another in time, and it is a matter of choice which one we find it most explanatory to specify.

[4] pp. 48–51.

Secondly, it enables us to avoid the paradoxes of continuity. Once the first point has been taken, and it is realised that one cause need not exclude there being other causes later in time, there is a temptation to analyse a cause into a chain of constituent causes. We explain how a cause results in its effect some ten minutes later by resolving it into a large number of little causal connexions; the big cause is merely the sum of the series of little ones, and to explain the big one is to specify the little ones. This procedure has the difficulty that if the big causes need to be explained in such a way, the little ones likewise require the same treatment, and there will be no end to the processes of subdivision,[5] unless it happens that there are atomic causes. If there are, then the causal schemata outlined earlier apply to those basic causal situations. But we do not know that there are, and there would be much the same difficulties as there would be with any granular theory of space and time.[6] For purposes of analysis, as well as for those of practical application, we need a concept of causality as a connected and continuous cord, which can be considered in as many steps as we please, but does not demand to be resolved into steps at all.

The outcome of the causal process thus depends not only on the initial causal situations but on the interval of time that has elapsed since. We take account of this by amending our functional dependences to be functions not of n arguments only but of $n+1$, n arguments representing the n initial conditions of the causal situation, and in addition a further one representing the temporal interval that has elapsed between the initial causal situation and the effect under consideration. We indicate the difference between the temporal parameter and the other conditions by separating it with a semi-colon instead of a comma; and we indicate the fact that it is not a date but an interval by writing not t but δt. The typical functional dependence is then of the form

$$Z_1 = f_1(a_1, a_2, \ldots a_n; \delta t).$$

Two further notational changes commend themselves. Cause and effect are both, now, specified by giving values—different values—to the same conditions. It is no longer helpful, now that the choice

[5] See, for example, Bertrand Russell, "On the Notion of Cause", *Proceedings of the Aristotelian Society*, xiii, 1912–13, pp. 1–26; reprinted in *Mysticism and Logic*, London, 1918, pp. 180–208. Pelican, 1953, pp. 179–96.

[6] See above, Ch. V, pp. 76–8.

of the date at which conditions are to be counted as initial conditions is arbitrary, to distinguish alphabetically the specification of the initial conditions from that of the resulting conditions. Instead of characterizing the former by the values $a_1, a_2, \ldots a_n$, and the latter by, say, $z_1, z_2, \ldots z_n$, we use the same variables to characterize them both, distinguishing the subsequent condition by the use of primes — say $x_1', x_2', \ldots x_n'$ — and the initial condition by the same variables unprimed — $x_1, x_2, \ldots, x_n$. A functional dependence is then expressed by a set of n equations in $n+1$ variables

$$x_1' = f_1(x_1, x_2, \ldots, x_n; \delta t)$$
$$x_2' = f_2(x_1, x_2, \ldots, x_n; \delta t)$$
$$. \qquad . \qquad .$$
$$x_n' = f_n(x_1, x_2, \ldots, x_n; \delta t)$$

Often we replace the function-variables $f_1, f_2, \ldots, f_n$ by $x_1', x_2', \ldots, x_n'$, and write

$$x_1' = x_1'(x_1, x_2, \ldots, x_n; \delta t)$$
$$x_2' = x_2'(x_1, x_2, \ldots, x_n; \delta t)$$
$$. \qquad . \qquad .$$
$$x_i' = x_i'(x_1, x_2, \ldots, x_n; \delta t)$$
$$. \qquad . \qquad .$$
$$x_n' = x_n'(x_1, x_2, \ldots, x_n; \delta t).$$

This notation is perspicuous, once it is realised that whereas in their occurrences on the left-hand side of the equals sign, $x_1', x_2', \ldots, x_i', \ldots, x_n'$, are variables whose values are real numbers, in their occurrences on the right hand side of the equals sign they stand for functions which we may or may not know, but which anyhow are not given here. On the right hand side we cannot talk of x_i' by itself, but only of the function $x_i'(x_1, x_2, \ldots, x_n; \delta t)$, which for example might be

$$2x_1 + \log x_2 - \tfrac{1}{2} x_n (\delta t)^2.$$

A functional dependence can be viewed in a number of different ways. It can be seen as a "difference equation", as the solution of a differential equation, as a one-one mapping or transformation of logical space — or phase-space as it is called — onto itself, or as a set of curves in a different phase-space. These different perspectives

reveal different assumptions about the nature of time and of causality, which, although interconnected, are not all the same, and need
different justifications.

Our explicit mention of the interval, δt, in a functional dependence
was to emphasize the fact that the magnitude of the interval between
the initial and final conditions is relevant, but the actual dates between which the interval lies is not. It does not matter *when* an initial
or final condition occurs, because laws of nature apply everywhere
and everywhen, and irrespective of personal attitudes.[7] But although
it does not matter *when* an initial or final condition occurs, it does
matter *how long* an interval elapses *between* the initial and final
conditions.[8] To put it another way, the dates themselves are
irrelevant, but the difference in dates is not: on the contrary, it is
highly relevant. Functional dependences therefore explicitly mention
differences of dates, but are not concerned with *absolute* dates. This
is not an analytic requirement. We could, conceivably, have laws of
nature which depended on the absolute date of the event, or the
interval that had elapsed since the big bang.[9] We have excluded such
a possibility by writing δt in the function. Often a bare t is written,
with the initial conditions being given the date t_0: such an expression
is useful for mathematical purposes, and is unobjectionable so long
as it is clear that t_0 is not a fixed constant but a variable one; that is,
a constant within the context of a particular calculation, but not
between one application of a functional dependence and another.
Mathematicians have no difficulty in handling it in this way, but for
the sake of clarity we continue, somewhat pedantically, to write δt
in order to emphasize that we are concerned with *differences* of dates.
Differences of dates occur explicitly in our calculations, although
they are time-invariant in the sense of applying to situations *at* any
date. A functional dependence is therefore, as regards time, in this
sense a "difference equation",[10] thereby expressing both the causal
irrelevance of temporal dates and the causal relevance of temporal
duration.

Since time is continuous, functional dependences can be, and often
are, expressed as differential equations. That is, instead of writing

---

[7] See above, Ch. I, pp. 10–12; Ch. II, pp. 22–3.

[8] See above, Ch. VIII, p. 119–20.

[9] See above, Ch. VIII, p. 126.

[10] The phrase 'difference equation' is not being used in its technical sense, where it
contrasts with a differential equation and signifies that the difference is not infinitesimal.

$x'_1 = x'_1(x_1, x_2, \ldots, x_n; \delta t)$

which is of the general form

$F(x'_1; x_1, x_2, \ldots, x_n; \delta t) = 0,$

we write

$G(x_1, x_2, \ldots, x_n; \dot{x}_1, \dot{x}_2, \ldots \dot{x}_n; \ddot{x}_1, \ddot{x}_2, \ldots, \ddot{x}_n; \ldots) = 0,$

that is, a function of $x_1, x_2, \ldots, x_n$ *and their time-derivatives*,[11] but not of a temporal variable itself. We can do this because, time being continuous, we can resolve a causal cord into steps that are as small as we please; and we often are impelled to do it, first because any choice of a non-infinitesimal difference is arbitrary, and secondly in order to accord with the principle of continuity. If time were discrete, we would naturally regard the situation at the instant next before an event as constituting *the* cause of that event in a privileged sense, although anterior situations could also be regarded as causes. By the same token, since time is not discrete, we regard the differential equation as expressing the real causal connexion. It alone satisfies Hume's requirement that the cause be temporally contiuous with the effect. Although the difference equation, in terms of some arbitrary non-infinitesimal interval δt, does also express the same physical law, it suggests that effect is, or at least might be, due to some temporally remote initial conditions without there being any intervening causal process. Once the initial conditions were satisfied, they mysteriously matured, and, when the time was ripe, issued in the effect, which sprang into existence like Athene out of the head of Zeus. The use of differential equations rules out any such suggestion, and shows that however unslowly the mills of God may grind, they grind in exceedingly small steps. Differential equations are thus peculiarly appropriate for expressing physical laws, satisfying as they do both the requirement of temporal repeatability and that of temporal continuity, and giving rise, when they are solved—if they can be solved—

[11] Following Newton, physicists often express time-derivatives thus:

$$\frac{dx_1}{dt} = \dot{x}_1, \quad \frac{dx_2}{dt} = \dot{x}_2, \text{ etc.}; \quad \frac{d^2x_1}{dt^2} = \ddot{x}_1, \text{ etc.}$$

to explicit equations which are independent of date but not independent of the duration elapsing between the initial conditions and the resulting conditions.

A functional dependence can also be viewed as the generalisation of the simple causal mapping of a discrete logical space, or phase space into itself. For any given value of δt, a set of equations in n other variables constitutes a mapping of some n-dimensional phase space into itself: granted certain further conditions,[12] it will be a one–one mapping of the space onto itself. For different values of δt the mappings will usually be different, and therefore we can regard them all as constituting a group of continuous transformations, transforming "points" $(x_1, x_2, \ldots, x_n)$ in phase space. It is a continuous one-parameter group, with δt as the parameter. For if we transform

$$(x_1, x_2, \ldots, x_n) \mapsto (x_1', x_2', \ldots, x_n')$$

and then

$$(x_1', x_2', \ldots, x_n') \mapsto (x_1'', x_2'', \ldots, x_n'')$$

we shall have transformed

$$(x_1, x_2, \ldots, x_n) \mapsto (x_1'', x_2'', \ldots, x_n'').$$

That is two transformations combine to form again a transformation. Also there is clearly an identity transformation, and, granted certain conditions, the transformations are reversible, so that given transformation T_a there is a transformation T_a^{-1}

$$(x_1', x_2', \ldots, x_n') \mapsto (x_1, x_2, \ldots, x_n).$$

Fourthly, if we have three transformations, T_a, T_b, T_c,

$$T_a(T_b T_c) = (T_a T_b)\, T_c.$$

That is, the transformations are associative. They therefore satisfy the four conditions for forming a group.[13] What distinguishes different members of the group are the different values assigned to δt. δt is thus a parameter of the group, and the only one. We therefore can apply a theorem about one-parameter continuous groups of

[12] See below, p. 174.
[13] See above, Ch. VII, p. 109.

transformations, that any such group is equivalent to a one-parameter group of *translations*. That is, there is a further transformation S, which will transform the transformations T of the group so that

if T_a is the transformation after an interval δt_a
and T_b is the transformation after an interval δt_b
and T_{a+b} is the transformation after an interval $\delta t_a + \delta t_b$
    then $ST_a ST_b = ST_{a+b}$.

In one way this is what we should expect in view of our principle of date-indifference. For if after an interval δt_a, $(x_1, x_2, \ldots, x_n)$ are transformed into $(x_1', x_2', \ldots, x_n')$ we then expect the same rule to apply for the development of $(x_1', x_2', \ldots, x_n')$ as for that of $(x_1, x_2, \ldots, x_n)$. Therefore we expect the transformation effected by the elapse of a *further* interval δt_b to be the same as would have been effected by an interval of equal magnitude, δt_b, but starting to run from the original date. What the theorem shows is that, so far as causal cords are concerned, date-indifference is not an extra requirement. Any group of transformations, continuous and in one parameter, can be retransformed into the desired form, where the result of the transformation over interval δt_a being followed by another transformation over interval δt_b is the same as the single transformation over interval $\delta t_a + \delta t_b$. So far as causal cords are concerned, we always can measure time so that temporal intervals are additive.

In another way it is a surprising result. It reveals a deep connexion between causality and time. It goes in the opposite direction to the two Hume noted,[14] of weak antecedence and contiuity, and is metrical rather than topological. If a causal law of development of physical systems is in some sense one-dimensional and continuous, and if it is transitive (which secures that the requirement of associativity is satisfied), and one-one, as required by Hume's rule 4,[15] then the causal process defines a metrical parameter with the properties of time. Weak antecedence is enough, granted the other requirements, to secure one-dimensionality. So we can always define a metric for time in terms of causal process. We have, in fact, already done so, in as much as we defined isochronous intervals in terms of periodic processes: but the present result is stronger, in that we no longer

14 Ch. III, p. 33.
15 Ch. IV, p. 44.

need the process to be periodic, and can rationally hope to be able to assign equal intervals to parts of a periodic process or to non-periodic processes. Given any causal process we can define a non-arbitrary metric for its parameter, and regard that as defining the measure of time. Many philosophers from Aristotle onwards have gone further and defined time itself in terms of causal process, but such a definition, although attractive on reductionist grounds, is open to objection.[16] Certainly the present argument does not warrant that conclusion, for it depends itself on time having the topological properties of continuity and one-dimensionality.

We can also consider functional dependences not as transformations in the n-dimensional phase space of conditions, but as curves in the $(n+1)$-dimensional phase space of conditions-and-time. It is this way of looking at it which gives rise to the picture of causal cords. Each set of conditions at a given date is a point, and the set of all possible sets of conditions at a given date is an n-dimensional hyperplane, being, as it were, a cross-section at that date. Each point in each cross-section is on one and only one causal cord. Each causal cord is one-valued with respect to time and never folds back on itself. It may, however, be periodic, and pass through the same point in the n-dimensional subspace of conditions a second time, and therefore an infinite number of times.

Thought

Perhaps we could adapt the simple ergodic argument for the discrete case, given at the beginning of this chapter (p. 158) to apply to the continuous case too. The discrete case turned on there being only a finite number of possibilities; no argument would be possible in the continuous case unless the n-dimensional "condition" space were finite, that is, the possible values of each condition were bounded. Even then, points and lines do not take up any room, and a causal cord could go through an infinite number of points in condition space without exhausting them all. But suppose we consider a small volume, containing all the points "near" a given point. Sooner or later, surely, the transforms of that volume would exhaust the whole

---

[16] See W. H. Newton-Smith, *The Structure of Time*, London, 1980, Ch. 2, especially Shoemaker's example, cited pp. 19–24; see also J. R. Lucas, *A Treatise on Time and Space*, London, 1973, § 2.

of condition space, and some part of the volume would be transformed back into the original volume. In which case a system must in the fullness of time return to *almost* its original condition. Or do we need to be sure that the way in which "volumes" of condition space are transformed will satisfy some further requirements?

We characteristically identify a causal cord by the point at which it intersects the first relevant cross-section—i.e. the initial conditions. Any one set of initial conditions picks out one system and its subsequent development through any given interval. The general statement of the functional dependence gives the rule of development for all systems: not a causal cord but a whole skein of causal cords filling the whole space, and constituting, so to speak, a causal rope, or bundle of fibres. We can then consider the whole space as filled with causal cords, which are not, as it were, followed through, but are considered as only indicating the direction of development at that point. To every point of our $(n + 1)$-space there is assigned a direction; that is to say, the ratios of $dx_1: dx_2 \ldots: dx_n: dt$, or, what is the same thing $\dot{x}_1: \dot{x}_2: \ldots: \dot{x}_n$. By the principle of date-indifference, we see that these are all functions of $x_1, x_2, \ldots, x_n$, but not of t. The directions assigned to each point of any cross-section will be the same for every cross-section. This feature is what is represented by the fact that when a natural law is expressed as a differential equation, the equation does not depend explicitly on t.

In the discrete case each condition either obtained or did not obtain, and the causal law could be expressed in truth-functional terms, and tested in accordance with the propositional calculus. Functional dependences are functions of continuous variables, which can take any of an infinite set of values, and cannot be expressed in simple truth-functional terms. They need to be tested, therefore, in different ways. A putative functional dependence can still be shown to be inconsistent with observed initial and resulting conditions, but it will never be possible to exhaust all possible combinations of relevant initial conditions, and instead we shall have to rely on assumptions of continuity[17] and simplicity, in order to establish its truth.

We consider first tests of relevance, and the analogue of the Method of Addition. We may have included in our list of possibly relevant variables some that are in fact totally irrelevant. This we

---

[17] Or sometimes, to be exact, differentiability. In all, except rather artificial cases, continuity implies differentiability.

show by having the corresponding equation take the degenerate form

$$x'_i = x_i$$

and all the other x'_1, x'_2, ... x'_n not depending on x_i at all. We can express the latter using the partial derivative

$$\frac{\partial x'_r}{\partial x_i}$$

which expresses how much x'_r would have been different if, in some given set of initial conditions and after a given period of time, x_i had been very slightly different, regarding the other variables as remaining constant meanwhile. We can arrange these partial derivatives in a table, thus:

$\dfrac{\partial x'_1}{\partial x_1}$ $\dfrac{\partial x'_1}{\partial x_2}$	$\dfrac{\partial x'_1}{\partial x_i}$	$\dfrac{\partial x'_1}{\partial x_n}$	
$\dfrac{\partial x'_2}{\partial x_1}$ $\dfrac{\partial x'_2}{\partial x_2}$	$\dfrac{\partial x'_2}{\partial x_i}$	$\dfrac{\partial x'_2}{\partial x_n}$	
$\dfrac{\partial x'_i}{\partial x_1}$ $\dfrac{\partial x'_i}{\partial x_2}$	$\dfrac{\partial x'_i}{\partial x_i}$	$\dfrac{\partial x'_i}{\partial x_n}$	
$\dfrac{\partial x'_n}{\partial x_1}$ $\dfrac{\partial x'_n}{\partial x_2}$	$\dfrac{\partial x'_n}{\partial x_i}$	$\dfrac{\partial x'_n}{\partial x_n}$	

In general, these terms will be functions of x_1, x_2, ... x_n; δt: but where one of the variables, say x_i, is causally irrelevant, the ith column will all be zeros, except in the ith row, where there will be a one. The reasoning is this: to say that, for all values of δt,

$$\frac{\partial x'_r}{\partial x_i} = 0 \text{ for all values of every } x'_r \text{ other than } x'_i,$$

is to say that no variation in the variable x_i will ever have any effect on any x'_r other than x'_i under any conditions, which means that the x'_r, other than x'_i itself, do not depend at all on x_i. For the sake of uniformity we reduce the first equation

$$x'_i = x_i$$

to a similar form, in terms of partial derivatives, namely

$$\frac{\partial x_i'}{\partial x_i} = 1.$$

Conversely, where one of the variables, say x_i', is not affected by the other variables, all the entries in the ith row will be zeros, except in the ith column, where there will be a one. For then, by similar reasoning,

$$\frac{\partial x_i'}{\partial x_r} = 0 \text{ for } x_r \text{ other than } x_i.$$

More strongly still, and more usual, where one variable neither affects the other variables nor is affected by them, both the ith column and the ith row will consist of zeros except where they intersect, where there will be a one. This can be expressed briefly, using the Kronecker delta,

$$\frac{\partial x_r'}{\partial x_i} = \frac{\partial x_i'}{\partial x_r} = \delta_{ir}.$$

To determine which of the possibly significant conditions is actually relevant, we vary each in turn, and see what, if any, difference it makes to each of the variable factors thereafter. This is, in effect, to determine what the partial differential coefficient of each of the variable factors is with respect to the varied factor, *under a particular set of conditions*. We have made a variation in the value of one of the variable factors, say x_r, and have increased it by, say δx_r, from x_r to $x_r + \delta x_r$. We have then compared the values of the variables that we had had with x_r, say $x_1', x_2', \ldots x_n'$, with the values of the variables we get with $x_r + \delta x_r$. The difference we write as $\delta x_1', \delta x_2', \ldots, \delta x_n'$; and, assuming continuity, the quotients

$$\frac{\delta x_1'}{\delta x_r}, \frac{\delta x_2'}{\delta x_r}, \ldots, \frac{\delta x_n'}{\delta x_r}$$

approximate towards the partial differential coefficients

$$\frac{\partial x_1'}{\partial x_r}, \frac{\partial x_2'}{\partial x_r}, \ldots, \frac{\partial x_n'}{\partial x_r}$$

as δx_r tends towards zero. If $\delta x_i'/\delta x_r$ is zero, within the limits of experimental error, we infer that $\partial x_i'/\partial x_r$ is zero for the given values of $x_1, x_2, \ldots, x_r, \ldots x_n$: δt, and say that x_r is irrelevant as far as x_i' is concerned. The exact justification of this is slightly different from the discrete two-valued case. There the elimination of irrelevant factors turned on our being able to factorise out the tautological $A_r \vee \overline{A_r}$. Here there is nothing analogous to that, but we follow instead the intuitively simpler procedure of subtracting the value of x'_i for the original x_j from its value for the altered $x_r + \delta x_r$ thus seeing exactly "what difference" the alteration from x_r to $x_r + \delta x_r$ makes. This is possible because with the real numbers subtraction is a well-defined operation. With the Boolean algebra of truth-values there is no corresponding operation of subtraction— $T \vee T$ yields T and so does $T \vee F$ —and therefore it was necessary to consider separately the Method of Difference and the Method of Addition. As in the discrete case, each new observation will show only whether a variation in a certain variable is causally significant or not *in a given set of circumstances*. Just as a certain factor may be causally irrelevant with respect to one combination of antecedent conditions but not to another, so $\partial x'_i / \partial x_r$ may be zero for some values of $x_1, x_2, \ldots, x_r,$ $\ldots x_n$, but not for all. In general $\partial x'_i / \partial x_r$ is a function of $x_1, x_2, \ldots,$ $x_r, \ldots x_n$, which may or may not vanish for certain values of the arguments.

This difficulty is, as we have noted, much more serious in the continuous case. For in the discrete case, with a finite and known set of possibly relevant factors, it was possible to test the relevance of any one factor against the background of any conjunction of other factors; for there were only 2^{n-1} different backgrounds. In the continuous case, however, even if there are only n possibly significant variables, each one varies over the whole or some part of the real number continuum, so that there are $2^{\aleph_0}$ (the cardinal of the continuum) of different possible values of the variables. It is thus impossible to consider them all one by one. But if all the functions are continuous, then we do not have to plot their value for every single set of values of the arguments, since the value of each function will not vary very much in the neighbourhood of each set of values. In strict mathematical terms, in order to determine completely a continuous function, we need to know the values for only a denumerable infinity of sets of arguments (provided they are dense in the reals), not for a whole continuum of sets of arguments. If we add to this some assumption of simplicity we can go very much further. We adopt the following procedure. We have found that $\delta x'_i / \delta x_r = 0$ for a certain set of values of $x_1, x_2, \ldots, x_n$: we then take a very much larger δx_r. If then $\delta x_i' \neq 0$, we know x_r is relevant to x_i' after all, and that $\partial x_i'/\partial x_r = 0$ only for the originally given value of x_j together with those of x_1, x_2, ... x_{r-1}, x_{r+1}, ... x_n. We were just passing through a maximum, minimum, or point of inflexion, of x_i' with respect to x_r, and our first spot was not typical. If, however, $\delta x_i' = 0$, we begin to believe that $\partial x_i'/\partial x_r = 0$ throughout the interval, and if we want further assurance, we take the mean value between the original x_r and the final $x_r + \delta x_r$, i.e. $x_r + \frac{1}{2}\delta x_r$, and if for that too the value of x_i' is the same, it is highly implausible that $\partial x_i'/\partial x_r$ is not zero throughout the interval—it would make x_i' a quartic in x_r; and even if so, one would be unlikely to have hit on three exceptional values. If x_i' appears not to vary, however we vary x_r under the given combination of the other conditions, we try altering them and again varying x_r. We make a random and fairly large alteration in x_1, x_2, ... x_n. Normally x_i' will be different. But if x_i' is unaltered, we begin to suspect that x_i' is independent of the other variables, and put this to the test by trying a few other spot values of x_1, x_2, ... x_n, and we then examine the effect of varying x_r, at first by a small amount, then by a larger, if a small variation has made no appreciable difference.

If we keep on getting $\delta x_i'/\delta x_r = 0$, we use the assumptions of continuity and simplicity to infer that $\partial x_i'/\partial x_r = 0$ under all conditions. For if not, then there would be some values of x_1, x_2, ... x_n; δt for which $(\partial x_i')/(\partial x_r) \neq 0$, and since the functions are all continuous, with continuous derivatives, the value of $\partial x_i'/\partial x_r$ will not fall away sharply to zero, but only gradually, and therefore at some of the values we had tried, $\partial x_i'/\partial x_r$ would not have vanished, even though it might be much less than at other values. But even if much less than elsewhere, it would still be detected when we made the large variation in x_r. Whereas in the discrete case, there is a serious possibility that we have overlooked one crucial factor—the key must fit the lock exactly, or else it will fit it not at all—in the continuous case everything is smoothed out, and if any factor has a serious effect at one point, it will most likely have some effect elsewhere. Although there are many—infinitely many—more cases, the cases need no longer be considered one by one; for the cases are connected, and what happens in one case will happen, though perhaps to a lesser extent, in other comparable, though not exactly identical, cases.

A *caveat* should be entered here. Not all natural ranges of factors vary in their causal efficacy continuously. There are critical temperatures and pressures. There are chemical processes which will take place within a narrow range of conditions but not at all otherwise. In particular, there are biological processes, where we find great independence of some variables combined with extreme sensitivity to others. Human beings will live under widely varying climates, and on widely differing diets, and will maintain the same internal temperature in spite of variations of external temperature (within limits), but cannot survive a 1 in 200 concentration of carbon monoxide. To express our understanding of biology we often need to reimpose the discrete binary schema of cause and effect upon the continuously variable substructure of functional dependence.

We may be able to find that out of our list of all the possibly significant variables, one, or more, is totally independent of the rest. This will, as we have seen,[18] show up in the table of partial derivatives as a row and a column of zeros with a one at their intersection, thus:

$$
\begin{array}{ccccccc}
\dfrac{\partial x'_1}{\partial x_1} & \dfrac{\partial x'_1}{\partial x_2} & \cdots & \dfrac{\partial x'_1}{\partial x_{i-1}} & 0 & \dfrac{\partial x'_1}{\partial x_{i+1}} & \cdots & \dfrac{\partial x'_1}{\partial x_n} \\[2ex]
\dfrac{\partial x'_2}{\partial x_1} & \dfrac{\partial x'_2}{\partial x_2} & \cdots & \dfrac{\partial x'_2}{\partial x_{i-1}} & 0 & \dfrac{\partial x'_2}{\partial x_{i+1}} & \cdots & \dfrac{\partial x'_2}{\partial x_n} \\[2ex]
& & & \cdots\cdots 0 \cdots\cdots & & & \\[1ex]
\dfrac{\partial x'_{i-1}}{\partial x_1} & \dfrac{\partial x'_{i-1}}{\partial x_2} & \cdots & \dfrac{\partial x'_{i-1}}{\partial x_{i-1}} & 0 & \dfrac{\partial x'_{i-1}}{\partial x_{i+1}} & \cdots & \dfrac{\partial x'_{i-1}}{\partial x_n} \\[2ex]
0 & 0 & \cdots & 0 & & 0 & \cdots & 0 \\[1ex]
\dfrac{\partial x'_{i+1}}{\partial x_1} & \dfrac{\partial x'_{i+1}}{\partial x_2} & \cdots & \dfrac{\partial x'_{i+1}}{\partial x_{i-1}} & 0 & \dfrac{\partial x'_{i+1}}{\partial x_{i+1}} & \cdots & \dfrac{\partial x'_{i+1}}{\partial x_n} \\[2ex]
& & & \cdots\cdots 0 \cdots\cdots & & & \\[1ex]
& & & \cdots\cdots 0 \cdots\cdots & & & \\[1ex]
\dfrac{\partial x'_n}{\partial x_1} & \dfrac{\partial x'_n}{\partial x_2} & \cdots & \dfrac{\partial x'_n}{\partial x_{i-1}} & 0 & \dfrac{\partial x'_n}{\partial x_{i+1}} & \cdots & \dfrac{\partial x'_n}{\partial x_n} .
\end{array}
$$

If we can thus eliminate all the irrelevant conditions, we are faced with the problem, to which there is no analogue in the discrete case, of finding the form of the functional dependence. There is no rule. It is usually a matter of inspired guess-work. We have an idea of the form the functional dependence should take, and juggle constants to try and find a formula of that form to fit the observed facts.

[18] p. 169.

Sometimes the formula does not fit very well, but is preferred on account of its simplicity to another that fits the data better. We think a linear equation is more likely to be true than a quadratic, a quadratic than a cubic; but often also prefer exponential and trigonometric functions to anything other than a linear one. If no formula suggests itself, we can use numerical methods to extract an approximation from the values of the partial differentials given by all those experiments where x_i was not irrelevant, and a variation in its value yielded concomitant variations in other variables: but such methods are cumbersome, and do not give much insight into the nature of the process being investigated.

Besides the simple case where experiment yields a single minimal set of variables which vary concomitantly with one another, there are two other possibilities we need to notice. There is first the possibility of the variables falling into two (or more) disjoint sets of mutually related variables. The variables of one set depend on one another, but do not depend on the variables of any other set. Thus it might be that the thermodynamic factors were completely independent of the electromagnetic factors and *vice versa*. If this were so, we could arrange the partial derivatives in the table to form two squares, with rectangles of zeros to fill up, thus:

$$
\begin{array}{cccccccc}
\dfrac{\partial x'_1}{\partial x_1} & \dfrac{\partial x'_2}{\partial x_1} & \cdots & \dfrac{\partial x'_m}{\partial x_1} & 0 & 0 & 0 & 0 \\[2ex]
\dfrac{\partial x'_1}{\partial x_2} & \dfrac{\partial x'_2}{\partial x_2} & \cdots & \dfrac{\partial x'_m}{\partial x_2} & 0 & 0 & 0 & 0 \\[2ex]
\cdots\cdots\cdots\cdots\cdots & & & & 0 & 0 & 0 & 0 \\[2ex]
\dfrac{\partial x'_1}{\partial x_m} & \dfrac{\partial x'_2}{\partial x_m} & \cdots & \dfrac{\partial x'_m}{\partial x_m} & 0 & 0 & 0 & 0 \\[2ex]
0 & 0 & 0 & 0 & \dfrac{\partial x'_{m+1}}{\partial x_{m+1}} & \dfrac{\partial x'_{m+2}}{\partial x_{m+1}} & \cdots & \dfrac{\partial x'_n}{\partial x_{m+1}} \\[2ex]
0 & 0 & 0 & 0 & \dfrac{\partial x'_{m+1}}{\partial x_{m+2}} & \dfrac{\partial x'_{m+2}}{\partial x_{m+2}} & \cdots & \dfrac{\partial x'_n}{\partial x_{m+2}} \\[2ex]
0 & 0 & 0 & 0 & \cdots\cdots\cdots\cdots\cdots & & & \\[2ex]
0 & 0 & 0 & 0 & \dfrac{\partial x'_{m+1}}{\partial x_n} & \dfrac{\partial x'_{m+2}}{\partial x_n} & \cdots & \dfrac{\partial x'_n}{\partial x_n}
\end{array}
$$

A second possibility is that we have too many variables, not because some are totally independent of others, but because they are totally dependent. An example of this would be the view put forward by John Locke, and later taken over by materialist philosophers, that all phenomena are functions of, and determined by, the "primary qualities" of the underlying substance; e.g. colour and sound depend on the structure of atoms and vibrations of molecules. On this view it would be unnecessary to list all the features of a system; only a subset—the primary qualities—need be listed, and the rest could be calculated from these. It turns out that we can determine whether a set of variable factors are independent or not by whether the "Jacobian" or "functional determinant"—that is the determinant formed of all the partial differential coefficients, thus:

$$\begin{vmatrix} \dfrac{\partial x_1'}{\partial x_1} & \dfrac{\partial x_2'}{\partial x_1} & \cdots & \dfrac{\partial x_n'}{\partial x_1} \\[2ex] \dfrac{\partial x_1'}{\partial x_2} & \dfrac{\partial x_2'}{\partial x_2} & \cdots & \dfrac{\partial x_n'}{\partial x_2} \\[2ex] \cdots & \cdots & \cdots & \cdots \\[2ex] \dfrac{\partial x_1'}{\partial x_n} & \dfrac{\partial x_2'}{\partial x_n} & \cdots & \dfrac{\partial x_n'}{\partial x_n} \end{vmatrix}$$

—is zero or not. Provided it does not vanish identically, the variables are all independent of one another; and, for any value of x_1, x_2, ... x_n and δt, provided it is not zero, the transformation is *reversible*, i.e. there are n n-adic, one-valued functions

$$x_1(\ldots), x_2(\ldots), \ldots x_n(\ldots),$$

such that

$$x_1 = x_1(x_1', x_2', \ldots, x_n')$$
$$x_2 = x_2(x_1', x_2', \ldots, x_n')$$

$$\cdot \qquad \cdot \qquad \cdot$$

$$x_n = x_n(x_1', x_2', \ldots, x_n').$$

The Jacobian thus provides a useful index for functional dependences. If $J \equiv 0$, then not all our variables are independent.[19] If

[19] As far as the mathematician is concerned, it does not matter which variables are regarded as depending on which. The distinction between primary and secondary qualities is not a mathematical one. If the secondary can be calculated from the

$J = 0$ or $J = \infty$ for some values of $x_1, x_2, \ldots x_n$, then there may be points of singularity where two or more causal cords unite or a causal cord diverges into two or more cords, causality, as we normally understand it, having temporarily broken down. But these are exceptional cases, and normally, for all values of $(x_1, x_2, \ldots, x_n)$, $J \neq 0$ and $J \neq \infty$.

We have now carried through the programme of generalising, in accordance with the principle of continuity, the discrete times and conditions considered by Hume in his analysis of causality into continuously varying magnitudes, whose causal interrelations can be expressed by functional dependences, and in particular, differential equations. We consider briefly how much further the programme of generalising from the finite and discrete to the continuous case can be pushed. Of the items set out at the beginning of this chapter we had already gone some way in Chapter IV towards generalising the third, where we had considered cases not only with a finite fixed number of possible causes but with indefinitely many. We can go further and generalise our *n*-dimensional spaces to Hilbert spaces, with a denumerable infinity of dimensions, so long as we are given them in some sort of order, so that we need not be hamstrung by our ignorance of some of the variables in the "tail". Nor need we be restricted to only a denumerable infinity of dimensions. A continuous function can be seen as yielding a set of co-ordinates, whose cardinality is that of the continuum. These generalisations are of great use in quantum mechanics. They also lead us in a different way to field theories in which, instead of considering the various factors of just one system, we consider the whole of space, comprising a non-denumerable infinity of points, each with causally relevant features assigned by some continuous function.

primary, then equally—provided some fairly standard conditions are satisfied—can the primary be calculated from the secondary. This is not to say the distinction between primary and secondary is groundless: but it is grounded on other than these purely mathematical considerations.

XI

Fields

CAUSES and effects, Hume said, should be contiguous.[1] We have taken Hume's contiguity to be really continuity, and have taken the requirement of continuity, or locality, not as an essential condition, but as only a *desideratum*. Given sufficiently good evidence of constant conjunction under all sorts of varying conditions, we should allow that a causal connexion held, even though we were unable to discover any spatio-temporally continuous link between them. Nevertheless we should be dissatisfied. We should feel that we were unable to give an adequate account of the causal connexion if we could not trace out the way the influence emanated from the cause and came to bear its influence on the effect. We might concede that it was established beyond reasonable doubt *that* the initial condition caused the subsequent condition, but reckon that we have not yet explained *how* the one caused the other: and we should not feel content until we had some idea—if only an outline explanation—of the "causal mechanism", the way in which the causal influence was propagated.

Locke weakened his requirement of continuity in the face of Newton's theory of gravitation.[2] Newton posited a gravitational force between masses that was proportional to their product and inversely proportional to the square of the distance between them, but could give no account of the way in which gravity operated, and assumed that it acted instantaneously. *Hypotheses non fingo*, he said: I do not pretend to offer explanations.[3] But this was a confession, not, as it was afterwards taken by positivists to be, a boast. He wrote to Bentley that it was "inconceivable that inanimate brute matter should, without the mediation of something else, which is not material, operate upon and affect other matter without mutual contact."[4] The something else was, he supposed, "an electric and elastic

---

[1] David Hume, *A Treatise on Human Nature*, Bk. I, Part III, Sect. XV, p. 173; see above Ch. III, p. 40.

[2] See above, Ch. III, p. 41.

[3] Isaac Newton, *Principia*, General Scholium, Alexander, p. 170.

[4] "Four letters from Sir Isaac Newton to Dr. Bentley, containing some arguments in Proof of a Deity", in *The Works of Richard Bentley*, III, London, 1838, p. 211.

spirit".[5] Clarke was equally clear that "nothing can any more act, or be acted upon, where it is not present",[6] and when Leibniz argued "'Tis a supernatural thing that bodies should attract one another at a distance, without intermediate means",[7] he acknowledged

That one body should attract another without any intermediate means, is indeed not a miracle, but a contradiction: for 'tis supposing something to act where it is not. But the means by which the two bodies attract each other, may be invisible and intangible, and of a different nature from mechanism; and yet, acting regularly and constantly may well be called natural; being much less wonderful than animal motion, which yet is never called a miracle.[8]

Leibniz, however, regarded an invisible, intangible, non-mechanical means of communication as inexplicable, unintelligible, precarious, groundless and unexampled. "But it is regular, (says the author), it is constant, and consequently natural. I answer; it cannot be regular without being reasonable: nor natural unless it can be explained by the natures of creatures."[9] Leibniz, essentially, was rejecting constant conjunction as an adequate criterion of there being a causal connexion, whereas Newton and Clarke reckoned that it could be an adequate criterion, although it did not constitute the whole meaning of the concept and we could not rest intellectually content until we had shown not only that it was regular but that it was indeed reasonable, that is to say explicable. If he had lived today, Leibniz would have been much sought after by tobacco companies to affirm that no causal connexion between smoking and cancer had been proved. As it was, in spite of his continual attacks on the chimerical, scholastic, occult quality of gravitation, since, together with Newton's three laws of motion, it fitted the observational facts, it was accepted as empirically established, and when electrostatic phenomena were investigated, Coulomb's inverse square law was accepted similarly.

The concept of a field was originally developed by Euler for hydrodynamics. Faraday made use of it to give an account of electromagnetic phenomena. His account was pictorial and intuitive. It was articulated into adequate mathematical form by Clerk Maxwell and

[5] Final paragraph of General Scholium, Alexander, p. 171.
[6] Clarke II, §4, Alexander, pp. 21–2.
[7] Leibniz, IV, §45, Alexander, p. 43.
[8] Clarke IV, §45, Alexander, p. 53.
[9] Leibniz V, §§121–2, Alexander, p. 94.

his successors. Fields have been generalised from being purely spatial to being spatio-temporal, and are used not only for theories of electromagnetism but throughout physics, and in particular to account for gravitation in the General Theory of Relativity. The concept of a field can, however, be best understood in simpler contexts, and we shall mostly consider in this chapter fields over space rather than fields over space-time, and simple gravitational, rather than complicated electromagnetic, ones. Feynman says that "a real field is a mathematical function we use for avoiding the idea of action at a distance",[10] but that is too simple. True, it is, as we have seen, a blemish on force theories that they involve the idea of action at a distance, and it is, correspondingly, a merit of field theories that they do not. But it is not their only merit. The concept of a field is a development of the line of thought among seventeenth century thinkers that regarded space as a plenum rather than a vacuum, a substance, to be referred to by a substantive, a noun, possibly with a capital letter, 'Space', rather than an attribute, merely indicating what place a particular thing occupied at a particular time. If space is a fundamental entity, a field is a way of telling us something important about it. It assigns to every point in the space a number—maybe a scalar, maybe a vector, maybe a tensor—and thus says something, perhaps something fairly complicated, about the space as a whole. Also, it generally satisfies (although we may allow exceptional points of singularity) the requirement of continuity, indeed, the slightly stronger one of differentiability. Thus, a field is a differentiable function from a space or space-time manifold into a set of abstract entities, which may be the set of non-negative real numbers, or the set of all real numbers. These are scalar fields. The abstract entities may also, as in Chapter VI, be vectors, which in an n-dimensional space can be represented by a set of n real numbers. Tensors are a further generalisation of vectors, and can, for a natural number m, greater than 1, be represented for an n-dimensional space, by an n^m array of real numbers. We can say that a field is a function on a continuous space

$$f: R^n \mapsto R^{(nm)}$$

for some non-negative integer m. If $m = 0$, it is a scalar field; if $m = 1$, it is a vector field; if $m \geqslant 2$, it is a tensor field.

---

[10]   *The Feynman Lectures*, II, §15.4.

We can, if we wish, use field theory for Newtonian gravitation. Instead of actual forces operating on actual masses, we associate with each actual mass a field of potential forces which would act on a "test" mass if it were there. That is, instead of saying that between any two actual masses, m_1 and m_2, there is a force from each to the other with magnitude

$$g \frac{m_1 \times m_2}{r^2}$$

where r is the distance between them and g a constant, we assign to m_1 a field of potential force, which at any point given by a vector $\mathbf{r}$ would exert on a test-particle of mass m_2, a force

$$-g \frac{m_1 \times m_2}{\mathbf{r} \cdot \mathbf{r}} \frac{\mathbf{r}}{\sqrt{\mathbf{r} \cdot \mathbf{r}}}$$

Thus far there is no great gain in using field theory, although there is a minor advantage due to the further fact that, granted a certain condition, a vector field, such as the one we have here for gravitational force, can be expressed as a "gradient" of a scalar field. Thus if on the (more or less) plane surface of the earth, I am given the gradient—the magnitude (e.g. 1 in 5), and the direction (e.g. ENE) of the steepest slope at that point—it is the same as if I were given its relative height. Obviously, given the latter—which is in effect a scalar field—I can determine the former. Less obviously, but still in fact provably, I can determine the latter from the former, granted a certain condition. The condition is that if one goes round any infinitesimal closed circuit—and hence if one goes round any closed circuit whatsoever—the total rise and fall due to the gradient around the whole circuit will be zero. Mathematically, this is called the "curl" of the vector field. It can be proved in vector calculus for any vector field $\mathbf{F}$ that

curl $\mathbf{F} = 0$

iff (i.e. if and only if) there is a scalar field V such that

$\mathbf{F}$ = gradient V.

The gradient of $\mathbf{V}$ is defined

$$\left(\frac{\partial V}{\partial x}, \frac{\partial V}{\partial y}, \frac{\partial V}{\partial z} \right),$$

as and is written grad V, or often ∇V. It is a vector. The curl of $\mathbf{F}$ in a three-dimensional space is defined as

$$\left(\frac{\partial F_z}{\partial y} - \frac{\partial F_y}{\partial z}, \frac{\partial F_x}{\partial z} - \frac{\partial F_z}{\partial x}, \frac{\partial F_y}{\partial x} - \frac{\partial F_x}{\partial y}\right),$$

where F_x, F_y, and F_z are the x-, y-, and z-components of $\mathbf{F}$; it is often written $\nabla \times \mathbf{F}$. It too is a vector.

Intuitively the condition makes sense. I cannot gain height on the surface of the earth by a circuit, no matter how small or large, that brings me back to where I start: my height depends only on my location, not on the path I have traversed. Equally in a gravitational field, the total amount of work I gain and expend in going round a closed circuit must be zero: I cannot extract work by going round in suitable circles; the gravitational field is, we say, "conservative". Hence, according to the theorem, we can view the field of potential force as the gradient of a scalar field—in fact the field of gravitational potential energy. This is an advantage, and yields some insight and some help in handling gravitational problems. But it is a minor advantage, and by itself would not be decisive. Other factors could be decisive, but do not, as it happens, apply to classical Newtonian gravitation. In electromagnetic theory it was a great advantage of field theory that it could handle lines of force which were not centrally directed and usually not straight. Again, the fact that the dielectric properties of the intervening medium are relevant in electrostatics is an empirical argument for fields being real. And when Clerk Maxwell developed a field theory that integrated electric and magnetic phenomena, and suggested further correlations with optics, and showed, in particular, that electromagnetic effects were propagated with the velocity of light, the arguments for the electromagnetic field were very strong. The gravitational field, by contrast, is extremely simple: there are no strong mathematical arguments for using field theory: no physical effects of the intervening medium have been detected: until the advent of the General Theory of Relativity, there was no connexion between gravitational field theory and any other branch of physics.[11] There was no evidence for gravitational effects being propagated with only a finite velocity. For these reasons, the issue between a field theory and an action-at-distance theory turned on philosophical considerations alone.

[11] For a full account of the ways in which the field theory of Newtonian gravitation differs from others, see J. C. Graves, *The Conceptual Foundations of Contemporary Relativity Theory*, Cambridge, Mass., 1971, Ch. 8, pp. 121–2.

There are, undoubtedly, some disadvantages in a field theory. It is likely to be mathematically indeterminate: although we may assign a gravitational potential to every point in space, we could equally well assign to every point one that was larger or smaller by the same amount. Again, fields do not give a good account of particles. Particles are seen as sources or sinks of fields, and as being acted on by fields, but within field theory itself appear as singularities, where the normal equations break down. Finally, although field theory satisfies very well the requirement of causal continuity, it does not satisfy at all well the continuity requirement of material identity. For point-particles, as for ordinary material objects, it is a necessary condition of something's being the same thing at two different times that it be spatio-temporally continuous between those two times, and a necessary and sufficient condition of their being different that they occupy different places at the same time. Fields occupy the whole of space, and different fields cannot occupy different places at the same time. Rather, if there are two fields, say caused by two different point sources, and granted that they are linear fields (as they usually are), they are "superposed" one on the other. The princple of superposition is characteristic of field theories as opposed to particle theories, which obey, rather, the principle of impenetrability, that one particle cannot be in the same place as another at the same time; mathematically, the principle of superposition is in many ways preferable, and easier to manipulate. But it merges the separate identities of the fields caused by different sources, and makes it impossible to trace a continuous identity for either one. To accept fields, therefore, is to abandon this important criterion of thing-hood. We may gain in explanatory power and in being able to account for the transmission of energy, but are losing the material entities that were preserved through all sorts of various changes.

These are real disadvantages, but not decisive ones. Nor are they the ones which have weighed with most thinkers, who have objected against fields their apparent insubstantiality, and the fact that we cannot see them or feel them. Empty space is insubstantial. Massy point-particles are much better candidates for being real, being idealisations of material objects, and the idea of their exerting forces on one another, however inexplicable, is intelligible—we know what forces are from our own experience of pulling and pushing. We only know what a field is by its effects. If there *were* a test particle at a point it *would* experience a force, or would accelerate. If we press the question 'How do we know?', fields turn out to be constitutionally "iffy"—mere potentialities not anything actual. They are open to the same objections as Berkeley made to the philosophers who gave a realist account of the tree in the quad. Although the language of field theory is categorical, and assigns to fields a firm ontological status, epistemologically all we find ourselves asserting are hypothetical claims about what would happen if certain experiments were done. Hence, it is argued, we ought, on grounds of both ontological economy and epistemological purity, to reject fields in favour of particles and forces.

The argument, however, is incoherent. It is incoherent to argue ontologically in favour of matter as the only reality, while adopting Berkeley's positivist arguments to show that we cannot ever know that fields exist. If Berkeley's arguments were cogent, they would tell as strongly against materialism—which was indeed Berkeley's original target—as against field theory. Epistemological arguments cannot be used against fields and in favour of point-particles, which are themselves as remote from sense experience as fields are. But if we are allowed to be realists with regard to point-particles, then we may find good reasons to be realists with regard to fields also. The claim that things are the only things to exist becomes an implausible dogma, once we realise that things can only exist through time and in space. Time and Space, and also Causality, are fundamental features of reality—as fundamental as things. Although we should not invent fields needlessly, there is no conclusive argument against their existence.

The counter-argument in favour of field theories that they treat Space, or Space-Time, as real entities with a juicy ontological status is two-edged. It fits the plenum view, but not so well the view that Space and Time are homogeneous and causally inefficacious. Of course, field theories attribute causal efficacy not to Space and Time themselves, but to fields, which are only attributes of Space and Time. Nevertheless, differences in spatio-temporal position may make a difference without any obvious difference in relation to material objects. The argument from continuity is far more telling. It goes with there being only a finite velocity of propagation of causal influence, and some degree of integration of space with time. If gravitational effects are not propagated instantaneously, we are more or less obliged to adopt a field theory: for suppose a hydrogen atom came into existence in the solar system; it would immediately be acted on by the gravitational force of the pre-existing sun, but the sun would be acted on by it only after some time, during which momentum would not be conserved. If, against this, it be maintained that hydrogen atoms cannot come into existence, we can argue along similar lines from rotating double stars which undoubtedly do exist. The only way of securing the conservation of momentum and energy, granted a finite velocity of propagation, is to attribute some momentum and some energy to the field. But then, if fields can have momentum and energy, they are strong candidates for being real substances, not mere unactualised non-entities.

The argument the other way is not quite so compelling. Hesse cites the example of the instantaneous transmission of pressure and longitudinal waves in an incompressible medium, which would be regarded as a continuous action.[12] But an absolutely incompressible medium is an idealisation. In all physically possible cases, a finite velocity of propagation is a necessary as well as a sufficient condition for continuous action, and if that is so, the requirement of spatio-temporal continuity leads us to field theories rather than action at a distance. This may provoke the further thought that the requirement of spatio-temporal continuity leads not only to a finite velocity of propagation, field theories, and an ontologically juicy view of Space and Time, but to some integration of, or at least to some integrated relation between, Space and Time, and hence the concept of Space-Time and the Special Theory of Relativity.

Further Reading

A M. B. Hesse, *Forces and Fields*, London, 1961, Ch. VIII, pp. 189–225.

B J. C. Graves, *The Conceptual Foundations of Contemporary Relativity Theory*, Cambridge, Mass., 1971, Ch. 8.

C Howard Stein, "On the Notion of Field", *Minnesota Studies in the Philosophy of Science*, V, Minneapolis, 1970, pp. 264–310.

[12] M. B. Hesse, *Forces and Fields*, London, 1961, p. 198.

XII

Continuity and Beyond

THE principle of continuity in its various forms has been of great importance, and in the last two chapters has led us to the concepts of a functional dependence and of a field, each of which leads us beyond the bounds of classical physics. In this chapter we shall consider ways in which the principle can be articulated and justified, and attempt to discern the outlines of the physics to which it leads.

The principle of continuity is partly a principle of locality, an anti-astrological canon. Remote factors are not immediate causes. They may be causes, but if so, they operate only through intermediate causes. There must be a continuous spatio-temporal connexion between cause and effect. An event can be caused only through events in its spatio-temporal locality. The principle is also a doctrine of the continuum, denying on the one hand the discreteness and on the other the gappiness of space and time. There is no next moment. Between any two instants of time or points of space there is another. Any division of instants into those that are all earlier than the later ones and those that are all later than the earlier ones, itself defines an instant; there is no partition of space into two disjoint closed subsets.

Locality is chiefly a requirement of explicability. It is also a concomitant of strong universalisability, and hence repeatability. Only if the influence of remote causes can be discounted, are we able to apply the concept of cause and test for possible causal laws.[1] But it is an epistemological prerequisite, not a logical consequence. Mach allowed the influence of remote causes, but held that experiments were repeatable. Mach also was a relationist, thus showing that a relationist doctrine of space and time does not entail any principle of locality. The principle of locality, however, entails some principles of repeatability and relationism. In saying that causes are mediated to their effects along spatio-temporal continuous paths we are saying that any effect depends on and only on the situation in its spatio-temporal neighbourhood, which is to say that if the situation in the spatio-temporal neighbourhood of one event is similar to that in the

---

[1] Ch. IV, pp. 57–61.

neighbourhood of another, the events must be similar too. We have some sort of repeatability. We also have a distinction between the situation *in* a neighbourhood and the spatio-temporal points themselves, and we are saying that events depend only on the former, and not on the latter. They do not depend on their absolute spatio-temporal position, and in maintaining the causal inefficacy of space and time, we are maintaining one relationist thesis.

The principle of locality is thus a powerful one, entailing, but not entailed by, the principles of repeatability and the causal inefficacy of space and time, although it needs to be presupposed in applications of the latter. It is, however, a difficult principle to express clearly. We do not easily form clear and distinct ideas of neighbourhoods. If Hume had been right, and the principle a principle of contiguity, we could have identified the immediate neighbourhood of any spatio-temporal point, and hence the situation in it: but granted continuity, there are no immediate neighbourhoods, and however close two distinct points are, a causal influence at one will have to be mediated through the intervening interval, if it is to yield an effect at the other. Neighbourhoods are infinitesimal neighbourhoods. These, however, we can deal with, thanks to the calculus. The infinitesimal neighbourhood is expressed in terms of dx, dy, dz and dt, or, when we wish to differentiate partially, ∂x, ∂y, ∂z, ∂t. In line with the use of differentials, we give the difference of the situation at the spatio-temporal point in question. Instead of saying that the effect depends on the situation in the infinitesimal neighbourhood, we say that the change in the situation at any one point depends on the way it is changing in the infinitesimal neighbourhood. We are thus led, once again, to talk of fields, only this time subject to the stronger conditions that the field satisfies a (reasonably simple) differential equation, and moreover, one in which x, y, z and t do not occur except as differentials. We are thus stipulating first that remote situations have no effect unless they are mediated through neighbouring situations, and secondly that only the neighbouring situation, and not any absolute value of spatial or temporal co-ordinates, is causally relevant—that is, that space and time are causally inefficacious.

Other conditions are often imposed. We normally require that the equations be similar as regards dx, dy, and dz, thus satisfying the isotropy of space. The simplest relation between the value of a scalar field at a point and its values in the neighbourhood is that the former should be equal to the average value of the latter, which can be expressed quite generally for an *n*-dimensional space by Laplace's equation

$$\frac{\partial^2 V}{\partial x_1^2} + \frac{\partial^2 V}{\partial x_2^2} + \ldots + \frac{\partial^2 V}{\partial x_m^2} = 0,$$

which can be expressed succinctly by saying that the divergence of the gradient of the scalar field V is zero, or in symbols

div grad V = 0,

or

$\mathbf{V}.\mathbf{V}V = 0,$

often written

$\mathbf{V}^2V = 0.$

If we take the four-dimensional case of Laplace's equation, and replace x_1 by x, x_2 by y, x_3 by z and x_4 by ict, where $i = \sqrt{-1}$, we obtain the wave equation

$$\frac{\partial^2 V}{\partial x^2} + \frac{\partial^2 V}{\partial y^2} + \frac{\partial^2 V}{\partial z^2} - \frac{1}{c^2}\frac{\partial^2 V}{\partial t^2} = 0,$$

in which the acceleration towards the average value in the neighbourhood is proportional to the amount by which it differs from it. The wave equation expresses the propagation of a disturbance in all directions with a finite velocity c. In a three-dimensional space it has a sharp front—in contrast to a two-dimensional space for example, where a stone thrown in a pond produces a sequence of ripples. Indeed, only in a three-dimensional space is it really possible to have all sorts of undistorted spherical waves,[2] so that if the propagation of causal influence is by waves travelling with a uniform finite velocity, we have an argument not only for integrating space and time into space-time, but for there being exactly three spatial dimensions of space-time.

The integration of space and time into a single space-time is accomplished by the Special Theory of Relativity. Of course, time is not the same as space, and any theory which said it was would be

[2] See R. Courant in E. F. Beckenbach, ed., *Modern Mathematics for the Engineer*, New York, 1956, p. 101; and R. Courant and D. Hilbert, *Methods of Mathematical Physics*, New York, 1962, vol. II, pp. 208–10, 688–91, 735–44, 763–6.

false. The Special Theory of Relativity works not with a four-dimensional space but with a (three-plus-one)-dimensional space-time in which there are important topological and metrical differences between space-like directions and time-like directions. Time is unlike space, but there is a sort of "trade-off" between time and space, represented by the uniform velocity of light, c. The partial similarity between time and space is reflected in a partial similarity between velocity and angle; a velocity says how much spatial distance I cover in a temporal interval, and an angle says how much spatial distance in one direction, e.g. North, I cover in transversing a spatial distance in another direction, e.g. East. Similarly there is a partial similarity between acceleration and angular velocity. Newtonian mechanics is covariant under a transformation from one co-ordinate system to another moving with a uniform velocity with respect to it, as also under a reorientation of the co-ordinate system, but is not covariant either under acceleration or angular velocity. These parallels seem much more natural in the Special Theory of Relativity. The Special Theory of Relativity also gives a deeper and more unified account of electricity and magnetism: it was for characterizing these phenomena that field theories were found to be indispensable. Historically, the Lorentz transformations, which are the characteristic transformations of the Special Theory of Relativity, were discovered as being those under which Maxwell's equations were covariant: but logically, Maxwell's equations may be seen as what, granted the conservation of charge, the equations must be, in order to be co-variant under the Lorentz transformations.[3] Magnetism is seen no longer as a separate phenomenon, but simply as how electrical phenomena must appear in a co-ordinate system moving at a uniform velocity with respect to another.

The Lorentz group of transformations is very like the Euclidean group, and the space-time of the Special Theory of Relativity is deeply Euclidean—indeed in a way more deeply Euclidean than the space of classical physics.[4] In both, the typical unforced motion is uniform motion along a straight line, and there is a sharp distinction between the geometrical background, which sets the stage where events happen, and the physics, which actually predicts or explains

[3] This approach is worked out in detail in E. M. Purcell, *Electricity and Magnetism*, (Berkeley, Physics Course 2), New York, 1965.

[4] A. A. Robb, *A Theory of Space and Time*, Cambridge, 1914. John A. Winnie, "The Causal Theory of Space-Time", *Minnesota Studies in the Philosophy of Science*, VIII, Minneapolis, 1977, esp. pp. 162–71.

the course of events. The General Theory of Relativity, by contrast, seeks to integrate geometry and physics, and to give an account of what happens without positing any general forces, such as gravity. It abandons the quasi-Euclidean structure of space-time, and instead of straight lines, considers geodesies—curves of minimum (or maximum) length—in a curved space-time. It requires a deep re-thinking of our concept of space and time, and in particular of causality, since there is a change of view on what counts as a natural motion, and what needs explaining by reference to a cause.

Of the four ways set out at the beginning of Chapter X in which the crude discrete concept of cause and effect might be generalised to a continuous analogue, we have in Chapters X and XI carried out three. It remains to consider the generalisation from the discrete truth-values T and F to the continuum of probabilities in the closed interval [0, 1]. Two branches of physics are probabilistic: thermodynamics and quantum mechanics. Not all the laws of thermodynamics and quantum mechanics are covariant under time reversal. Time there seems to have a direction, which it does not appear to have in classical and relativistic physics. And whereas it is easy, given an asymmetric relation, to define a symmetric one, viz. the union of the asymmetric relation and its converse, it is difficult, given a symmetric relation, to define an asymmetric one. This suggests that the account of time given in classical and relativistic physics is only a derivative one, obtained from something deeper by leaving out the feature of directedness.

Thermodynamics can, but quantum mechanics cannot, be viewed simply as the statistical treatment of a large number of individually determinist systems. Quantum mechanics appears—although this is contested—to be ineliminably probabilistic. But the introduction of a continuum of probabilities in place of discrete truth-values appears to introduce *discontinuities* not only into the concept of causality but into that of substance too. For we use continuity not only to identify but to distinguish. In the $(n+1)$-dimensional picture of a causal process outlined in Chapter X, each curve shows both how different states can be states of the same system—if they both lie on one and the same causal cord—and how different states must be states of different systems—if they do not lie on the same causal cord. This is a discrete question, admitting only of the answers Yes or No. So, too, when we use spatio-temporal continuity to individuate substances. Either a a thing is the same as some thing observed on a different occasion, or it is not. Once, however, we introduce probability, we no longer have causal cords, but only causal "vapour trails", and there will be no reason why the result should be one particular outcome rather than another. And whereas the subjects of ordinary true or false propositions can reasonably be taken to be substances, probabilities apply basically not to propositions but to propositional functions, whose subject is not a definite term referring to a substance, but a variable whose reference is deeply unclear. The programme of everywhere generalising from the discrete to the continuous case thus runs into difficulties, and these difficulties appear to be not merely contingent ones, due solely to the fact that quantum mechanics happens to be the theory which best fits the empirical facts, but, at least partly, conceptual ones as well—we have already seen how,[5] in a very different field, out of the continuous groups of translations and re-orientations there emerged the discrete group of reflections and parity reversals. It seems that our intimations of discreteness and atomicity are derived as much from metaphysical principle as from physical fact.

[5] Ch. IX, p. 153.

Further Reading

Although there are many philosophical problems still in classical physics, the reader is likely to want to move beyond the limits of this book. The Special Theory of Relativity is difficult but not impossible, and the philosophically inclined reader is likely to find most useful

E. F. Taylor and J. A. Wheeler, *Spacetime Physics*, Freeman, 1963.

Roger B. Angel, *Relativity: The Theory and its Philosophy*, Pergamon, 1980.

A number of other authors are writing on the philosophy of physics, on similar lines to this book but dealing with more advanced topics. I have seen and discussed early drafts of their work, and shall myself be looking out for:

H. R. Harré and I. J. R. Aitchison, *The Philosophy of Special Relativity and Electromagnetism.*

Robert Weingard, *Lectures on General Relativity and Philosophy.*

M. L. G. Redhead, *Incompleteness, Nonlocality and Realism—A Prolegomenon to the Philosophy of Quantum Mechanics.*

Peter Gibbins, *Particles and Paradox.*

M. J. Lockwood, *Realism, Determinism and the Quantum Universe.*

Appendix on Relationism

Notes towards answers to questions at end of Chapter IX

(1) *Is it Obvious that it would Make no Difference if Everything were Doubled in Size?*

No, it is not obvious, because it has not been made clear what else might have been changed too. Suppose first that nothing else whatever had been changed, and in particular, that time, as measured by a caesium clock, was unaltered. Then we should find that the velocity of light was only 93,000 "new miles" a second, and it took about 5 seconds to send a radar pulse to the moon and back. But perhaps time intervals might be doubled too. What about mass? We might say that masses had remained the same, with a corresponding reduction of density, or we might say that densities had remained the same with masses being increased eightfold. If masses remain the same, then gravity at the surface of the earth will diminish fourfold, because according to Newton's law of gravitation

$$g = k \times \frac{M}{r^2}$$

where k is the gravitational constant, M, the mass of the earth, is assumed to be constant, and r is assumed to have doubled. This will affect the period of a pendulum, which is $2\pi\sqrt{l/g}$. Not only will l be doubled, but g will be only a quarter of what it was, so the period of a pendulum will be only $\frac{1}{2}\sqrt{2}$ of what it was in old seconds, that is $1/\sqrt{2}$ of what it should be in new seconds, and radar pulses to the moon and back will take less than two new seconds as measured by a pendulum clock. If, however, densities remained the same, the mass of the earth will be increased eightfold, and hence g will be doubled. In that case the pendulum clock will go on ticking in old seconds rather than new seconds, and we shall once again find the velocity of light has halved.

Other suggestions may be made. But clearly if the value of c, the velocity of light, and h, Planck's constant (which determines the duration of the vibration of the caesium atom) remain unaltered, some discrepancy will emerge somewhere, whatever additional assumptions are made.

Galileo was the first to argue that the size of things could not be greatly altered if the rest of physics were to remain unchanged, in the second day of *The Discourses and Demonstrations Concerning Two New Sciences*. A modern argument is given by G. Schlesinger, *Philosophical Studies*, vol. XV, no. 5, October 1964, pp. 65–71, and countered by A. Grünbaum in the same volume, pp. 71–9.

(2) *Different Versions of Relationism*

    (i) A space, properly so called, is not just a set of one or more points, but a set together with a *structure*, that is to say a system of relations holding between the members of the set.

(ii)   (a) We cannot properly speaking either give individual names or ascribe specific properties to points or instants, but only state relations holding between them.

(b) Although we may be able to give individual names or ascribe specific properties to points or instants, we can translate without loss into relational terms.

(c) Although there could conceivably be a loss in translating into relational terms, in fact there is no loss of empirical truth in so doing.

(d) Although there could conceivably be some loss of empirical truth in translating into relational terms, there could not be any loss of physical truth.

(e) Although there could be some loss of physical truth, in fact there is not.

(iii)   (a) Only local spatial and temporal relations are of interest to physicists.

(b) Global properties of space and time as a whole can be translated without loss in terms of local relations.

(c), (d), (e) as in (ii).

(iv)   (a) Only actual material objects and events exist, and only actual spatial or temporal relations between actual material objects or events are of interest in physics.

(b) Only actual or possible material objects and events can exist, and only actual or possible spatial or temporal relations between actual or possible material objects or events can be of interest in physics.

(v)   (a) Only actual material objects and events, together with actual spatial or temporal relations between them, are perceived, and all knowledge must be of them alone.

(b) Only actual or possible material objects and events, together with actual or possible spatial or temporal relations between them, can be observed, and all intelligible discourse must be of them alone.

(vi)   (a) As a matter of empirical fact, physical laws are covariant under (α) spatial translation (β) temporal translation (γ) spatial reorientation (δ) spatial reflection (ϵ) temporal reflection (ζ) uniform linear velocity (η) angular velocity (θ) acceleration (ι) angular acceleration (κ) any diffeomorphism (that is, any transformation that preserves differential relations) (λ) magnification.

(b) As a matter of methodological principle, physical laws ought to be covariant (α)–(λ).

(c) As a matter of conceptual necessity all universal laws of nature must be covariant under (α)–(λ).

(vii)   (a) As a matter of empirical fact, there is no physical way of distinguishing a preferred frame of reference with respect to (α)–(λ).

(b) As a matter of methodological principle, there should not be a physical way of distinguishing a preferred frame of reference with respect to (α)–(λ).

(c) As a matter of conceptual necessity, there cannot be any rational way of distinguishing a preferred frame of reference with respect to (α)–(λ).

In this classification Newton would be a relationist in the sense of (ii) (d), (vi) (a) (α)–(ζ), (vii) (a) (α)–(ζ). Leibniz would be a relationist in the sense of (ii) (a), (iii) (a), (vi) (c) (α)–(ζ) and (vii) (c) (α)–(ζ). Mach would be a relationist in the sense of (v) (b), (vi) (c) (α)–(ι) and (vii) (c) (α)–(ι), being ready to invoke whatever additional hypotheses that are necessary to accommodate the phenomena. Einstein maintained (vi) (b) (κ), and hence (vi) (b) (α)–(ι) and (vii) (b) (α)–(ι), in his principle of equivalence, being prepared to make great alterations to the concept of space and in the geometry he ascribed to space-time in order to secure that physics should be the same whatever the frame of reference employed.

It is easy to expand the list of relationist doctrines, and to introduce further refinements in the ones already given. In many cases it is difficult to be sure where a particular thinker stands on some questions—including some of those given in the previous paragraph. Few thinkers answer questionnaires for the benefit of posterity. They are led to their conclusions by various considerations, and their conclusions both are formulated and are to be interpreted in the light of the arguments that led to their acceptance. Possible positions, in the absence of grounds for adopting them, are empty of philosophical interest. Hence the feelings of exhaustion and irritation the reader will have experienced as he worked through the seemingly exhaustive list given here. Nevertheless, he may find it a useful exercise to make his own exhaustive list, after having considered arguments for and against relationism.

(3) *Different Arguments for Relationism*

(i) Things—*i.e.* material objects—are the only things to exist, and events occurring to them are the only events to happen. Unoccupied positions and uneventful instants are nonentities, and cannot be talked about, save as the scene for material objects and events occurring to them.

(ii) Things and events occurring to them are the only things that can be observed. Unoccupied positions and uneventful instants cannot be known about except through positions occupied by material objects and instants when events occur.

(iii) Although other things than material objects—*e.g.* rainbows—or events—*e.g.* the music of the spheres—might be observed, the public identifiability and re-identifiability of material objects and their associated events is so central to our conceptual scheme of any possibility of there being a public language, that material objects must occupy a privileged position in our conceptual scheme, and all science must be expressible in terms of them and relations between them.

(iv) Whatever other things exist, empty space and eventless time do not, being defined in terms of non-existence.

(v) Whatever other things can be observed, empty space and eventless time cannot.

(vi) Causes must be repeatable. A mere difference of space or time cannot make any difference *per se*. Space and time are causally inefficacious; that is to say, space and time must be homogeneous, and space must be isotropic, in which case only relations between points, and not any monadic quality of any individual point, will be invariant or of any causal significance.

(vii) A respectable scientific explanation does not refer to mere position or mere orientation in space or to mere date in time, but only to distances, angles or durations.

(viii) Only if it can be expressed as a differential equation in which spatial and temporal co-ordinates do not occur explicitly, will a generalisation be regarded as a really respectable law of nature.

(ix) Physics is not concerned with the particular but the general, and therefore only with laws that are covariant under a large group of transformations.

(4) *Different Arguments against Relationism*

(i) From the fact that a space must have a (relational) structure, it does not follow that no point in it can have any non-relational property. The natural numbers have order-type ω, but the number One has properties all of its own; in the real number continuum R, with order-type θ, the zero has unique properties.

(ii) We need to talk about Space and Time as a whole in order to discuss their global topological properties—their dimensionality and whether they are continuous, unbounded, non-cyclic, orientable.

(iii) Although spatial positions and durations and temporal dates cannot be observed directly, they may be determined indirectly—as non-rotating frames of reference are, and a date for the beginning or end of the universe might be.

(iv) It is a matter of empirical fact whether there is a preferred frame of reference. Although, as it happens, Newtonian mechanics, the Special Theory of Relativity and the General Theory of Relativity are covariant under wide groups of transformations, it is logically possible that the physical theory which best fitted the phenomena was covariant only under an extremely narrow group of transformations. Hence there can be no conceptual arguments for relationism, but only an empirical determination of the degree of homogeneity and isotropy that our universe happens to have.

(v) Although the homogeneity and isotropy of space and time may be as much a *fiat* as a fact, and may be argued for as a conceptual necessity or stipulated as part of the definition of physics, it cannot be a requirement of covariance in respect of absolutely every transformation, or there would not be any *laws* of physics at all. There must be some limits to the group under which laws are to be covariant, and therefore some features, such as acceleration in Newtonian mechanics, that are absolute.

(vi) Relationism, if it is to be plausible, must talk about the possible as well as actual, and it is much more objectionable to talk about unrealised possibilities than unoccupied positions or uneventful instants.

(vii) Even if we allow talk about possibilities, it is unclear what sort of possibilities are envisaged. Logical possibilities are too wide, physical possibilities too narrow, to give an adequate characterization of space, while if we talk about the spatially possible, or the temporally possible, we seem merely to have replaced a noun by an adverb.

(viii) Even if we can give a non-circular characterization of space and time in terms of possibilities, we seem only to be *describing* spatial and temporal phenomena; while if we allow that Space and Time are real, we may be able also to give an explanation.

Index

absolute
  a. angular velocity 139
  a. dates 162
  a. orientation 138
  a. position 138, 142
  a. space 122, 138, 146, 149
  a. spatio-temporal position 185
  a. time 122
  a. value 185
  a. zero in space 126
  a. zero in temperature 90, 102
absolutely zero 142
acceleration 133, 139, 141, 187
  angular a. 140
action
  a. at a distance 41, 58, 178, 180, 183,
    natural unit of a. 102
addition 54
  a. rule for magnitudes 93
  a. rule 95, 96, 97, 102
aesthetic beauty 81
agent 25, 66, 128
agents 35, 36
Alfred's candles 98
allotrope 16
alteration 116
alternative explanation 48
  alternatives (in another world) 129
analytic 29, 30, 66, 69
analytical truth 3, 16
Angel, R. B. 189
angle 97, 187
  apparent a. 112
angular
  a. acceleration 140
  a. momentum 99, 118
  a. velocity 110n., 133, 139, 140, 141,
    187
animism 10
antecedent 42
  weakly a. 67
  weak antecedence 33, 36, 37, 165
antisymmetric 72
apparent angle 112
application of concepts 27, 28, 29, 33
arbitrary 128
Archbishop of Canterbury 8, 9
Aristotle 107

Aristotelian conceptions of space 104
associative law 109
astrology 60, 136
  astrological 57, 184
asymmetric 72, 87
  asymmetry 144, 145
atom 39
attribute (space as an a.) 148, 149, 178,
  182
axes 129
Ayer, A. J. 22

balance 85
  balancing 86
beginning (of time or space) 82
Bennett, J. 144, 145, 154
Berkeley, G. 123n., 142, 182
big bang 134, 162
binary (logic) 156
biology 10, 65, 107
Bleen 20
Boolean algebra 170
Bose-Einstein
  B-E particles 131
  B-E statistics 131
Bostock, D. vi, 104
Bosworth, R. C. 150n.
boundary conditions 7, 128
Broad, C. D. 142
brute (facts) 12, 21, 60, 117, 127
bucket experiment 132, 133, 134, 135,
  136, 137, 138, 142, 143

caesium 97, 99
Buridan's ass 129
calorie 92, 94
Cantor 76
Carroll, Lewis 154
Cartesian demon 65
  Cartesian doubt 27, 63, 67
Cathedral clock 64, 65
causal
  c. inefficacy 137, 138, 182, 185
  c. irrelevance 55, 56, 57
  c. laws 24, 82
  c. mechanism 176
  c. necessity 33
  c. relation 18, 21

causality 13, 14, 18, 27, 69, 137, 165,
    188
causation 15
cause 19, 27, 55, 138, 150
    c. of change 138
centrifugal force 132, 133, 136, 138,
    139
chance concomitances 36
change 138
charge 83, 102
Charles' Law 92
chemistry 10, 11, 106
China 60
cigar 46, 47, 52, 53, 54
Clarke 104, 122, 129, 132, 133, 143,
    148, 177
class-inclusion 89
classification 107
clocks 64
closed universal 34, 36
cobalt 70
Cocchiarella 78n.
coefficient of expansion 91
coincidence 32, 36, 63, 65, 66
Collingwood, R. G. 37, 42, 55
colour 107, 111, 113, 173
combination 48, 49, 50, 52, 54
communication 111
comparatives 72
concept
    application of concepts 27, 28, 29, 33
    conceptual 123
concussion 24
conditions 128
    boundary c. 7, 128
conditiones sine qua non 55
conjunctive normal forms 50
connected 88
    c. temporal antecedence 72f.
conservation 67
    c. of momentum and energy 183
conservative 180
constant conjunction 31, 32, 34, 36, 40,
    41, 44, 63, 67, 71, 105, 106, 128,
    176
contiguous 40, 44, 96
    contiguity 41, 176, 185
contingent 3, 16, 34n., 61, 65, 141
    contingency 127, 128, 129
continuity 41, 71, 124, 156, 163, 166,
    167, 171, 174, 176, 178, 182, 184,
    185
continuous 42, 119, 188

c. function 170
c. qualitative concepts 67
c. transformations 143
continuously variable substructure
    172
continuum 73, 77, 78
contiuity 33, 41, 57, 67, 163, 165
contradiction 31
'contravariant' 116n.
convention 98
co-ordinates 83, 104, 124, 125
Coulomb's inverse square law 177
counter-example 54
Courant, R. 186n.
covariance 82, 114
    covariant 114, 116
curl of vector field 179
cyclic 71, 82, 158

dates 83, 162
date-indifference 165, 167
Davidson, D. 68n.
Davies, P. C. W. 105
Dedekind 74, 76
deductive 5, 6, 14, 17, 18, 31, 48, 63
    d. justification 16
    d. logic 30, 33
degrees (of temperature) 91
Denbigh, K. A. 150n.
dense 73, 75, 76, 77
    d. space 78
Descartes 9, 27, 35
determinism 65
    determinist 66
diagrams 24
diamond 16
difference 116, 117
    d. between weights 95
    "d. equation" 162
    d. of position 59, 126
different 26, 107, 111, 125
    'different from' 106
differentiability 167n., 178
differential equations 82, 156, 162,
    175
dimension 71, 175
    d. number 78
    d. of time 82
    d. of space 78, 124, 146
direction of time 82
discontinuous transformations 143
discrete 73, 75, 76, 156, 171, 184, 188
    d. binary schema 172

d. causality 159
d. orderings 77
d. qualitative concepts 67
d. space 78
d. structure of space and time 79
disjunction 49
disjunctive normal form 49, 157
displacement 110, 111, 120, 125, 141
dissimilar 107, 114
'dissimilar' 106
distance 102
distant 59
Dorling, J. 150n.
doubled 120
dualism 153
duration 97, 162

Earman, J. 141n., 142, 154
East 127, 150
Eberhard, P. H. 66n.
economy 7
effect 31, 67, 157, 158
egocentricity 22, 23
Einstein 2, 4, 58, 81, 122, 136
E-Podolski-Rosen Paradox 58n.
Einstein, Mrs 2
electromagnetic factors 173
  e. phenomena 177
  e. theory 180
  electromagnetism 178
electron 103
elegance 7, 20, 81
eliminate 45
  elimination 48, 51, 54, 172
  elimination by addition 64
  eliminative induction 18
Ellis, B. 104
empirical
  e. content (of law of conservation of
    mass and energy) 117
  e. evidence 79
empiricism 3, 7, 81, 128
  empiricist 137
  the Empiricists 27, 28
end of time 82
endeavour 21
Enderton, H. B. 72n., 87n.
energy 114, 117, 183
  kinetic e. 94
entity (space as an e.)
  non-e. 123, 183
equality of differences 93
equivalence

e. class 85, 86, 89, 131
e. relations 72, 86, 88, 100
ergodic theorem 158
Erlangen Programme 110n.
Erlickson, H. 58n.
error 84
ethologist 107
Euclid 131n.
  Euclidean group 110, 120, 148, 187
  E. metric 149
  E. space 90
  proper E. group 110, 120
Euler 58, 143, 177
evidence 26, 70, 80
exist 19, 25
expansion
  coefficient of e. 91
experience 1, 4, 20, 23, 28, 41, 42, 69
experiment 8, 21, 22, 35, 36, 60, 61, 64,
    66, 127, 132
  bucket e. 133, 134, 143
  experimental facts 20
  experimenter 65, 66
  e.'s interventions 60
explain 40, 129, 150
explanation 38, 127, 137, 140, 141
  alternative e. 48
explicability 184
  explicable 33, 40, 177
extensionalist 33, 38
external world 19, 20, 69

fact 12, 30, 126, 138
  brute facts 12, 21, 60, 117, 127
false 80
  falsifiability 5, 7, 18
  falsified 44
  falsifying 32
Faraday 177
Feynman 178
fiat 126, 131, 138
field 184
  field theories 175
finite 74
  f. velocity of propagation 182, 183,
    186
fire 30, 113
First Antimony of Pure Reason 74
first-order mechanics 140
flat 28, 35, 36
Fonda, L. 154
force 138, 141, 178, 179
four-vectors (4-vectors) 81

function 157
  continuous f. 170
function dependence 161
functional dependence 82, 128, 156, 159, 160, 162, 166, 167, 172, 175, 184
functional dependency 108
functional determinant 174
future, the 24, 25, 26, 29, 150
  future events 14, 15, 24
Galileo 99
  Galilean group of transformations 131, 134
  Galilean physics 104
gappiness 184
Gardner, M. 142, 151, 154
general 15, 23
  g. explanations 59
  generalization 14, 36
General Theory of relativity 100, 116, 122, 126, 138, 139, 149, 178, 180, 188
geodesics 188
geometry of appearances 112
Gestalt psychologists 24
Ghirandi, G. C. 154
Gibson, J. J. 112n.
globes 132
God 127, 128, 133, 158
Goodman, N. 20
gradient 179
granular
  g. structure of time and space 76, 78, 160
graphite 16
Graves, J. C. 13, 142, 180, 183
gravitation 4, 58, 176
  Newtonian g. 179, 180
  gravitational effects 182
  gravitational field 180
  gravitational forces 138
  gravity 176, 188
Greeks, the 107
Gregory, R. S. 112n.
group 111, 117
  g. theory 110, 114
Grue 20
Grunbaum, A. 150n.

habit 28, 69
harmony 1, 5, 80, 81
Harré, H. R. vi
Hart, H. L. A. 55n.

'here' 22
Hesse, M. B. 183
hexagon 132, 145
Hilbert spaces 175
Hill, E. L. 150n.
historians 105
homogeneity of time 117
homogeneous 71, 120, 182
homomorphism 85, 88
Honoré, A. M. 55n.
Hooker, C. 142
human
  h. affairs 37, 38
  h. being 57
  h. knowledge 129
humane 59
humanities 12
Hume, D. 28, 30, 31, 33, 34, 37, 38, 40, 41, 42, 44, 67, 69, 75, 82, 105, 106, 107, 157, 158, 163, 165, 175, 176, 185
Huntington, E. V. 73n., 82
Huygens 99
hyperbolic tangent 101
hypothesis 45, 48, 49, 54, 105

'I' 22
ideal gas 92
ideas 27
identity 71, 109, 130
  i. in difference 110
  i. transformation 108
  i. of indiscernibles 39, 117, 127, 129, 130, 131, 132
ideology of science 10
impenetrability 21, 181
impersonal 22
impossible 34
  impossibility 21
inclusion 102
incongruous counterprats 143, 146, 149
inconsistency 65
indexical 22
indicative 35, 36
individuate 71, 126
induction 14
inductive
  i. arguments 5, 16, 17, 27, 48
  i. inference 13
  i. policy 25
  inductively 70
inefficacy 119, 126
  causal i. 137, 138, 185

causally inefficacious 182
ineluctability (of laws of nature) 20, 21, 114
inertial frame 104
infinite
 i. class 74
 i. potentiality of human beings 38, 39
information 150, 158
initial conditions 140, 141, 159
inscrutable 128
instance 40, 45, 54, 62, 105, 106, 107, 129, 141
integers 73, 83
intersubjectivity 113
interval 162
interventions 66
"inus" 54
invariance 67, 110, 114, 116
 temporal i. 113
 invariant 78, 109, 111, 112, 114, 115, 116, 117, 125
inverse 109
 i. square law 177
 i. transformation 108
inversion 143
irreflexive 72, 87
irrelevance 51, 57, 60, 61, 65, 67, 70, 71, 97, 106, 119, 162, 172
isochronous 97, 121, 165
isomorphic orderings 74
 isomorphism 85
isotope 16, 39
isotropic 120
 i. space 78
 isotropy 117, 139, 185

Jacobian determinant 174
Jammer, M. 58n., 146n.
Jerusalem (as natural zero) 104
Johnson, Dr 21

Kant 24, 28, 69, 74, 143–9, 154
Kelvin 92
Kilmister, C. W. 142
kinds 18
kinetic energy 94
 kinetic theory of gases 92
Klein, Felix 110n.
Kneale, W. C. 82n.

language 23, 24, 107
Laplace 186
latitude 84

lattice 121
law 19
 causal l. 24, 82
 natural l. 7, 18, 20, 22, 25, 36
 l. of nature 12, 21, 27
left 144, 145
Leibniz 34, 39, 57, 58, 104, 117, 122, 127, 129, 131, 132, 135, 138, 143, 145, 150, 177
 L's law 130
 Leibnizian monads 141
Lemmon, E. J. 72n.
length 97, 102
levitate 8, 9
'like' 106
limits 90
linear order 88
lines of force 180
 straight lines 90, 187, 188
locality 41, 57, 58, 119, 184
Locke, J. 27, 41, 48, 111, 173, 176
Lockwood, M. J. 101n.
'logical' 30
 logical necessity 30, 31, 34
 Logical Positivists, the 28
 logical space 157, 161, 164
 logicality 80, 81
longitude 84
Lorentz 149, 150
 L. transformations 187

Mach 58, 122, 134, 135n., 136, 137, 139, 184
Mackie, J. L. 43, 44, 54n., 68n.
Magic Cube 107, 109
magnetic field 70, 126, 139
 m. pole 83
magnifications 110
magnitude 91, 93, 95
 addition rule for magnitudes 93
mapping 157
mass 95, 103, 114, 117, 125, 126, 179
 negative m. 90
massergy 117
material entities 181
 material objects 19, 21, 25, 113, 181
materialism 173, 182
Mathematical Induction 77
Maxwell, C. 150, 177, 180
 M's equations 187
mean kinetic energy 92
meaning 29
measurement 67, 82

mental attitude 59
Mercator 98
metaphysical 82
method of addition 51, 55, 167, 170
Method of Agreement 62
Method of Difference 61, 62, 170
metrical 165
Michelson–Morley experiment 4, 7, 105
Mill, J. S. 44, 61, 62
Miller 7
Milne, E. A. 99n.
minimum perceptible size 75
minimum sufficient condition 54, 55, 67
miracles 8, 10
Möbius strip 146, 147
modal logic 33
Möhr 83, 90
momentum 118, 183
  angular m. 99, 118
Moses 30
motion 120, 138

Nagel, Ernest 114, 142
natural law 7, 18, 20, 22, 25, 36
  natural necessity 36
  natural numbers 73
  natural zero 90, 104
nature
  laws of nature 12, 21, 27
necessary 21, 30, 33, 42
  'necessary' 34
  logically necessary 33
necessity 12, 33
  causal n. 33
  logical n. 30, 33, 34
  natural n. 36
negation 49
  negative conditions 63
  negative entropy 150
  negative factors 48
  negative mass 90
neighbourhood 119, 185
Nerlich, Graham 142n., 146n., 147n., 153
Neumann, Peter M. 118
Newton 41, 58, 122, 127, 133, 134, 135, 136, 137, 139, 140, 149, 150, 163n., 176, 177
  N's first law 140
  N's second law 140
Newtonian

N. conceptions of space and time 132
N. gravitation 179, 180
N. mechanics 95, 99, 138, 139, 141, 150, 187
N. physics 104
Newton-Smith, W. H. vi, 72n., 79n., 81n., 82, 154, 166n.
non-entity 123, 183
non-orientable space 146, 147
non-subjectivity 20
normal 138
  conjunctive n. forms 50
'now' 22
nth order mechanics 141
numerically distinct 71

objective
  o. features of the world 149
  o. shape 111
  objectivity 84, 113
objects 21
observation 127
  o. statements 12
observers 36
obstacles 21
Ohm 94
omnitemporal 15
one-dimensionality 166
ontology 148
  ontological 19, 21, 22, 114, 130, 131, 182
open statements 34
open texture 63
open universal propositions 36
open-ended claim 29
optics 180
order 87, 88, 91
  strict order 87, 89
  ordering 72, 88
  order-types 73, 74, 76, 89
ordinal requirements 124
orientable space 71, 82, 147
orientation
  absolute o. 138
origin 129
  o. of ideas 27
  o. of time-scale 134
  natural o. in space 104
original sin 23
oysters 45, 46, 47, 52, 53, 54

Pantin, C. F. A. 11
parapsychology 9, 10
parity 70, 143, 151

part 89, 138
  part–whole relationship 89, 90, 96
particles 181
  Bose–Einstein p. 131
particular 15, 21, 23, 28, 29, 31, 178
  particularity 105
partition 86
Paschal full moon 2
past (the) 25, 26, 36, 150
  'past' 22
  past events 14, 15
Pears, D. F. 154
penny 111, 112
perception 111
perfect symmetry 132
periodic processes 165
  periodicity 121
person 22
perspective 116
phase-space 161, 164
Phenomenal Regression to the Real
  Object, the 112
phenomenalism 8, 12, 19, 21, 23, 66,
  67, 112, 148
phlogiston 90
phosphorus 16
photons 131
physical intuition 80
physics 10
  Newtonian p. 104
place 21, 22
Planck's constant 102
Plato 1, 2, 3, 11, 12, 13, 18, 22, 105
  Platonic universals 21
plausibility 7
plenum 178, 182
Podolski 58
Popper, K. R. 5, 6, 7, 45, 150n.
port 45, 47, 52, 53, 54
positivists 130
'possible' 34
  possibility 114
  possible objects 148
  possible spatial relations 148
pragmatic 25
predictions 25
pre-established harmony 63, 64, 65, 66
present, the 150
  'present' 22
pride 23
primary qualities 111, 113, 173
principle of sufficient reason 37, 127,
  129

Prior, A. N. 78
probabilities 188
probe 21, 36
projection 157
proper Euclidean group 110, 111, 120,
  131
proton 126
pseudo-angle 101
psi phenomena 9
psychokinesis 70
Purcell, E. M. 187n.

qualitatively identical 71, 106
quantum mechanics 41, 58, 79, 81, 99,
  117, 118, 131, 188
Quine, W. V. 79n.
quotient set 87, 88

radical empiricism 3, 4, 8, 12, 80, 81
rapidity 101
rational 20, 23, 25, 32, 40, 80
  'rational' 30
  r. agents 39
  r. choice 129
  r. numbers 73, 74, 76, 77, 96
rationalism 2, 12, 80, 81
  rationalist approach 118
  Rationalists, the 3, 27, 28
rationality 11, 13, 20, 21, 22, 23, 32, 38,
  128
real 12, 21, 181, 183
  r. numbers 76, 77
realism 80, 81, 112, 182
reality 21, 80, 114, 182
reason 5, 12, 28, 41, 42, 69, 128
  reasonable 177
Redhead, M. L. G. 58n., 118
reductionism 166
reflected 120
  reflection 110, 143, 151
reflexive 72, 86, 87
regraduate (a magnitude) 101
regularities 128
Reichenbach, H. 26, 142
relation, space as a 71, 148, 149
  causal relation 18, 21
  relational 124, 126
  relational structure 72
  relationist 122, 184
  'relative' 122n.
relativity
  see General Theory of r.
    Special Theory of r.

relevance
  causally relevant 55, 56, 67, 162
Remnant, P. 144, 154
remote 57, 60, 136, 184, 185
  remoteness 137
reorientation 110, 111, 120, 141
repeat 141
repeatability 31, 36, 37, 38, 40, 44, 67, 119, 163, 184, 185
repeatable 9, 30, 33, 42, 71, 84
'resembling' 106
resistance 93
restricted science 11
reversible time 150
rhythm 121
right hand 144, 145
rigid bodies 97
rigid movement 144, 148
rigid rulers 100
Robb, A. A. 187n.
Rosen, S. P. 58, 153n.
'rotation' 110n.
Rubik's Magic Cube 107, 109
rulers 96, 98
Russell, B. 14, 42, 57, 156, 160

St. Mary's clock 64, 65
Salmon, W. C. 26, 68n., 83
same 26, 111, 114, 115
  sameness 106, 110, 125
scalar field 179
scale 86, 91
sceptic 17, 22, 23, 24, 28, 34, 48, 59
scepticism 25, 47, 67
Schlege, Richard 150n.
Sciama, D. W. 58, 142
Second Law of Thermodynamics 158
second-order mechanics 139
secondary qualities 113
self-contradictory 15, 17, 30
self-sufficiency 23
sense-data 8, 12, 19, 24, 69
sense experience 8, 19, 28, 67, 114, 182
sensorium 133
shape 110, 111, 112, 113
Shoemaker, S. 166n.
shoulders (standing on other men's s.) 61
similar 106, 107, 114, 115, 116
simple enumeration 62
simple order 88
simplicity 7, 20, 81, 167, 170, 171, 172

simply connected space 71, 82, 147
simultaneity 81
Singmaster, D. 105
'sinister' 153
size 95, 110, 111
Sklar, L. 118, 134, 141n., 142, 147n., 148, 154
Smart, J. J. C. 154
solipsism 23
sorts 18
sound 173
space 13, 21, 27, 42, 57, 59, 72, 90, 122, 123, 137, 146, 147, 149, 178
  Aristotelian space 72
  space as an attribute 148, 149, 178, 182
  Newtonian conceptions of space 132
  non-orientable space 146, 147
  orientable space 71, 82, 147
  topological properties of space 78, 146, 147, 148
space–time 78, 149
spatial orientation 59
  spatial position 59
Special Theory of Relativity 4, 7, 20, 73, 78, 80, 81, 90, 99, 100, 101, 102, 122, 134, 136, 183, 186, 187
spell 10, 37, 51, 70
square of opposition 34
Stapp, H. P. 58
stars 134, 135
states of mind 65
Stein, H. 134, 142, 183
Stoll, R. R. 72, 87n.
Stoy, G. A. 118
straight line 90, 187, 188
Strawson, P. F. 17
strict ordering 87, 88, 89
subgroup (of transformations) 109
subjective 112
  s. error 85
substance (space as a s.) 121, 146, 148, 149, 178
substantiality 12
subtraction 170
sulphur 16
superposition 181
Suppes, Patrick 72n.
swans 14, 15, 17, 19
Swinburne, Richard 14, 26, 81n.
symmetric 72, 86, 87
symmetry 71, 82, 116, 118, 128, 132
symmetry group 119, 141, 142

synthetic assumptions 70
synthetic *a priori* propositions 69
synthetic statements 3, 16

taxonomy 107
Taylor, E. F. 101n., 189
temperature 19, 92
temporal invariance 113
temporal position 59
tense 22
tenseless 22
tensors 20, 83
theorem 33
theoretical statements 12
theory
   t-loaded 83
   t of groups 107
thermodynamics 92, 173, 188
thermometer 91
'thin' 28
thing 19, 21, 120, 123, 182
thing-hood 181
third-order mechanics 139, 140
'this' 22
Thompson, E. C. 118
Thouless, R. H. 112n.
tides 99
time 13, 21, 22, 27, 42, 57, 59, 74, 122,
   123, 137, 149, 165
   Newtonian conceptions of time 132
token-reflexive 22
Tolman, R. C. 5, 80
tomorrow 14
topological 165
   topological invariant 71
   topological property 82, 111
   topological properties of space 78,
      146, 147, 148
   topological questions
   topological requirements 124
topology 78, 102
transformation 107, 108, 109, 110, 157
   continuous transformation 143
transitive 86, 87
   transitive temporal relations 72
'translation' 110n.
trial and error 35
trial and failure 21
triangle of opposition 34
true 80
   truth 79, 98, 106, 122
   analytic truth 3, 16
   truth-table 49

truth-values 156, 170
try 25
types 18, 40, 45, 61, 105, 106

ugly functions 20
unbounded space 82
underdetermination 79, 80, 82
uniform velocity 141
uniformity of nature 15, 18, 21, 116
   uniformity in the world 23, 24
universal 19, 29, 38, 105
   closed universal 34, 36
   universal laws 5, 7, 11
   Platonic universals 21
universalisability 184
   universalisable 29, 59, 38
universality 12, 21, 34, 67
unlike 26
   'unlike' 106
unreal 121

vacuum 123, 178
vectors 83
velocity 100, 102, 187
   angular velocity 110n., 133, 139,
      140, 141, 187
   velocity of light 100, 180
   uniform velocity 141
verification 28
   verification principle 130
   verificationist 133, 142, 144, 145
Vyaltzev, A. N. 78

water 30, 31, 114
Watson, C. J. H. vi
wave equation 186
weak antecedence 33, 36, 37, 165
   weakly antecedent 67
weight 86, 95, 96
West 127, 150
wet 114
Weyl, H. 118
Wheeler, J. A. 101n., 189
Whitrow, G. J. 99n., 150
whole 89, 138
Wigner, E. P. 71n., 118, 128n., 143
will 21, 113, 134
Winnie, J. A. 187n.
Wittgenstein, L. 146n.
Wright, G. H. von. 44n., 67n., 78n.
wrong 19

Zangwill, O. L. 112n.
zero 90, 91, 102
  absolutely zero 142
  zero-order mechanics 141

η order-type η 74, 76
θ order-type θ 76
τ-time 99
ω^* order-type ω^* 74, 75